Satellite Projects Handbook

Dedication

This book is dedicated to the memory of my late father, Ronald, who always waited patiently for the NOAA 10 evening satellite pass to finish, before asking me to drive him home.

Satellite Projects Handbook

Lawrence Harris

NEWNES

Newnes
An imprint of Butterworth-Heinemann
Linacre House, Jordan Hill, Oxford OX2 8DP
A division of Reed Educational and Professional Publishing Ltd

    A member of the Reed Elsevier plc group

OXFORD   BOSTON   JOHANNESBURG
MELBOURNE   NEW DELHI   SINGAPORE

First published 1996

© Lawrence Harris 1996

British Library Cataloguing in Publication Data
A catalogue record for this book is available from the British Library

ISBN 0 7506 2406 X

Library of Congress Cataloguing in Publication Data
A catalogue record for this book is available from the Library of Congress

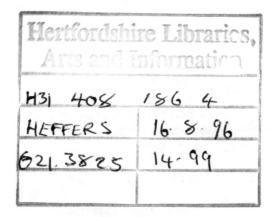
Typeset by Vision Typesetting, Manchester
Printed by Hartnolls, Bodmin, Cornwall

Contents

Preface

The purpose of this book is to explain why we launch satellites, what they do, and how to take part in this exciting world without a huge financial outlay. I hope also to convey the sense of wonder and tension experienced when a satellite is launched, having taken an active role as a satellite controller.

Early chapters include summaries of the operations of successful British satellites, in order to illustrate the strong relationship between professional and amateur operations. The main part of this book concerns those satellites to which you can easily tune your receiver (that is, those from which you can receive telemetry), and the fascinating projects that are possible.

Scientific, weather and other research satellites are featured, together with examples of the types of data or pictures that they transmit. The costs currently required to assemble suitable receiving equipment, and the various types of hardware available, are discussed.

Recent years have seen a surge of interest in the computer decoding of satellite data, so information is given on this important aspect, including the opportunity to obtain some useful software. With new products coming onto the market almost monthly, details are given to help the reader to decide on the most suitable current software and hardware for satellite work.

The role of satellite operations in the school curriculum is discussed. A number of schools currently operate satellite receiving stations. Because of the tremendous educational potential, this subject merits much wider involvement. A number of projects are described for the benefit of staff who might wish to use satellite data for such purposes.

Included in the book are references for additional information on products, contact addresses etc., but the book is intended to be complete in its coverage of the subject, and I hope to induce in the reader an enthusiasm for satellites, and a desire to take part in this fascinating field.

Acknowledgements

A number of people and institutions have provided help during the preparation of this book:

- Andy Hancock for his help on various occasions
- Dave Cawley of Timestep Weather Systems for letting me have first try of new hardware some years back
- Des Watson (formerly of the RIG committee) who provided me with Kepler elements before I went on-line
- European Space Agency, Publications Division for permission to quote from their documents where appropriate; the periodical *Image* is published by EUMETSAT, which provided information on future METEOSAT operations
- Geoffrey Falworth, the editor of the specialist magazine *Satellite News* who provided me with many of the Kepler elements that were difficult to obtain, and helped to identify several mystery transmissions
- Geoffrey Perry and Les Currington for permission to refer to their specialized work with Russian satellites
- Goddard Space Flight Centre for the provision of Kepler elements, and helpful booklets
- Joe Jones of Plymouth Radio Club who provided so much help with my framestore
- Matra Marconi for permission to reproduce pictures of ARIEL-6, a satellite with which I was closely involved
- NASA and its associated departments who provide the satellites to which we all listen, and for the several publications that they release to keep the large body of amateurs satisfied
- NOAA Technical Memorandum NESS 95 for information on NOAA satellites
- Timothy (my son) who made various special electronic circuits for me, and converted a BBC predictions program for the (then) new PC

- World Meteorological Organisation who provided a number of diagrams such as WMO411 for inclusion
- Paul Wilson who sent me disks full of space news
- Dr Paul Williams, Director of the Rutherford Appleton Laboratories for permission to publish information on various satellite projects
- Duncan Enright, Sadie McClelland and Renata Corbani at Butterworth-Heinemann for making this book possible
- And lastly, but not least, my wife Marion for countless cups of tea and support–and hours spent preparing diagrams and helping to check readability.

Computer software

Satellite tracking software and Kepler elements

Several good tracking programs are available from various sources within the weather satellite fraternity. Those wanting to obtain the latest copy of PCTrack can send the author their name and address and enclose five pounds sterling. In return you can expect to receive a disk containing the latest PC version of this shareware program, together with the latest set of Kepler elements for a large number of satellites, including the weather satellites. This package enables you to be up and running within a few minutes. Using this software, together with a conventional receiver, should enable you to positively identify dozens of satellites transmitting in the 137 MHz, 149 MHz and other bands.

PCTrack is a multi-satellite tracking program for the PC, which holds a database of up to 300 satellites and can display the results in different projections. Details for registering the software are included in the extensive documentation which comes with the program. The registered version is an enhanced program permitting greater control of the display.

Write to:
Lawrence Harris
5 Burnham Park Road
Peverell
Plymouth
PL3 5QB
UK

Computer software

Satellite-tracking software and Kepler elements

1 Operations with UK-5, UK-6 and IRAS

This chapter has two purposes: the first, to provide probably the only published record of the operations work on three specific, major British space projects; the second, to enable the reader to understand the background to working with satellites. Those contemplating making a career in the space industry may wish to familiarize themselves with as many aspects of satellite operations as possible. This chapter covers the day-to-day activities of such operations.

Various professional satellite operations described here have their equivalent in amateur applications. Facilities offered by new generations of computers enable the hobbyist to perform what was once an impossibly expensive process for the amateur to consider. You will appreciate the effort that goes into any space project, and this text merely hints at the dedication of the staff involved.

This work was carried out by the Satellite Control Centre, initially at Datchet (near Slough), as part of the programme of research funded by the Science and Engineering Research Council. This Centre formed part of the world famous Radio and Space Research Station, later renamed the Appleton Laboratory. Laboratory staff (and projects) were later transferred to the Rutherford Laboratory in Oxfordshire.

Professional satellite operations on UK-5 (ARIEL-5)

My involvement with this started when I entered the 'Satellite Operations' group (which had responsibility for day-to-day running of the Satellite Control Centre), as a member of the support team. This was during the mid-1970s, when the team was running and maintaining software used to process data.

The Control Centre team consisted of operations staff – three teams on shift work – and support staff who ran software for administration and data processing. A management team of senior staff oversaw the project. We start with a look at the UK-5 satellite operations, UK-5 being primarily a satellite containing equipment to monitor X-rays coming from space.

Satellite predictions

There is an obvious need to calculate the times at which a satellite will become visible from your ground station. The UK-5 satellite project did not have a tracking facility of its own; we used some NASA stations, mainly Quito (in Equador) and Ascension Island (in the South Atlantic). The latter was used as a back-up station. There was therefore a requirement to calculate when these stations would see the satellite. In return for this valuable help, the American science teams had full access to the resulting data.

One of the many benefits of space research has been agreement between nations to share scientific results, often on an arrangement such as that of UK-5. Space projects are expensive – though this should be carefully balanced against many other considerations – and it is always in the interests of nations to cooperate in such ventures.

Data collection

The process of collecting data was as follows: the Control Centre would run programs on the mainframe computer to calculate those times during the day when the satellite would pass over Quito. This 'pass schedule' was then transmitted to NASA headquarters using teleprinter circuits, in the form of a request for the use of their Quito station. In due course, NASA would schedule Quito wherever possible, or alternatively Ascension Island if Quito was not available. Occasionally neither was available.

NASA has control of many satellites, ranging from interplanetary missions, near-Earth orbiting and manned missions, to cooperative ventures such as UK-5. Naturally many of these had a far greater priority than UK-5, so sometimes we had to miss out – i.e. not collect data from a pass. Rarely did we do so though! Scientists and administrators can work well together, and we often went to extreme lengths to avoid missing data. On many occasions, a helpful American would let us borrow his 'voice' line for long enough to allow us to use it for a data transmission!

Voice and data links

Queries were dealt with using the SCAMA service – a telephone connection via London Switching Centre – that linked the whole of the NASA network. There were many occasions when I had to call London and ask to speak to a NASA department, followed, after a short period while the connection was made, by my talking to a native of Quito about satellite data. One should

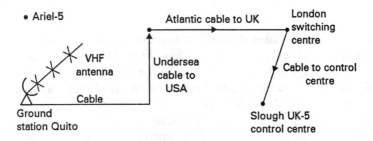

Figure 1.1 *Data flow from UK-5 (ARIEL-5) to British Control Centre*

appreciate the marvel of being able to speak to an Ecuadorian within a few seconds, using a combination of ground lines to London, undersea lines to NASA, or perhaps the occasional satellite communication link. High-technology of the late 1970s at its best!

Some of the lines linking the Control Centre to the London Switching Centre were designed for voice communications and others for data. They crossed the ocean via communications satellite or direct undersea line to NASA, then on to the ground station itself. There were normally two separate lines set up (we use the term 'configured') – one for voice and one for data. Data was usually satellite telemetry (the signal transmitted by the satellite). On more than one occasion, due to line shortages, we had to drop the voice line and do the operation blind! This all added to the interest and very rarely did we lose data.

As you might expect, the various lines were set up some twenty minutes before each pass was due to start. This period was used to review details of the pass between the ground station and the control centre, using a printed summary called a pass briefing message – or 'PBM'. Each side could then be sure that both knew the exact procedures to be followed. There were also contingency arrangements to be carried out in the event of one or both lines going down (failing) – just in case a special command sequence was being carried out. This did happen on occasions.

During the pre-pass checks, commands to be transmitted to the spacecraft were sent first to the ground station, so that they could be verified as correct. This transmission was done using a PDP-8 mini-computer (1978). The station would take just a few minutes to check these commands (which contained built-in checksums), and were then stored ready for the pass. It was not unknown for the verification check to fail, leaving some important decisions to be made in just seconds! During this check, the ground station transmits simulated satellite data back to our Control Centre, in order to monitor the line quality. Following the command 'configure for AOS (acquisition of signal)'; this simulated data would be removed a few minutes before the real pass was due to start.

AOS!

When the ground station (Quito) first acquired the satellite signal, the station operator would announce 'Quito has AOS UK-5'. Data would then flow – intermittently to start with, due to the very low angle at acquisition. Early data is always rather poor, due to the effects of the atmosphere on the signal coming from the satellite. You will see this yourself, should you decide to set up a receiving station. Within a minute or so, as the satellite's elevation increases, the signal improves and the operator announces 'Quito has solid lock' – meaning that there are no drop-outs in the signal. The term 'lock' refers to the receiver circuits which lock on to the signal. Data now being received by the ground station is transmitted along data lines, following the route previously described, until it enters the control centre computer. In this machine (another PDP-8), it was then analysed by hardware and converted to digital format, then recorded on tape.

Housekeeping

During the first few minutes of a typical pass, the incoming telemetry is analysed by the computer in real time (live) and all of the systems on the satellite are checked. As an example of an important check, the voltage across the battery terminals on UK-5 was always monitored, particularly towards the end of the satellite's life, to ensure that it was high enough. When the satellite entered eclipse, this voltage would drop dramatically and continue to fall. Just before leaving eclipse, it would be at its lowest. There was concern that if the voltage dropped too quickly, it might cut off power to some of the systems, requiring major operational effort to restore the experiments to their normal states. In such an instance, it is preferable to turn off one experiment rather than risk a battery cut-out.

This procedure of checking out the systems on-board a spacecraft is called 'housekeeping', and in a later chapter, you will read of its relevance in your own satellite work.

So after routine checks during a typical pass, the only commanding to be done would be to 'dump' the stored data – data that has been stored on the satellite's tape recorder, or in its computer's memory, while it orbits the planet. One command is sent – causing the satellite telemetry to change from real time to 'dump' data. Dumping this stored data took just a few minutes and then the telemetry reverted to real time.

At that time, back in the mid-1970s, the domestic microcomputer was merely a dream. Our programs were usually run on a mainframe computer, an ICL 2960 costing many thousands of pounds. Software that does essentially the same job has been written to run on a home computer costing a small fraction of the cost of those original machines. Software to process real time and other data, from certain British satellites (to be detailed later), is now available for several machines, so you can take part in the most interesting application of computer technology that can be devised for a home microcomputer.

The bulk of the scientific data, (we called it the 'bulk'), was processed on an ICL 2960. To input the huge amounts of data obtained each day, a PDP-8 mini-computer, capable of handling large volumes of data, was used. Data was stored on high-quality digital tape and, in a later chapter, you will learn how to store your own satellite data on rather similar tape, if of somewhat lower quality!

UK-6 (ARIEL-6)

The next, and sadly final, satellite in the successful UK series was the cosmic ray satellite UK-6, launched in 1979 (Figure 1.2). This became the first satellite to be controlled by a British ground station, that at Winkfield in Berkshire. Originally a NASA station, Winkfield was being phased out of duties with the American network. For our British team, including me, the UK-6 operations at Winkfield were an exciting development that allowed us to become very closely involved with the tracking itself, rather than commanding the spacecraft remotely from Datchet.

Figure 1.2 *The British scientific satellite ARIEL-6*

Before the launch of UK-6, the training phase allowed us all to visit Winkfield, and many new friendships were made. It was also during operations with UK-6 that I was promoted to Satellite Controller (shift leader). This was the first satellite for which I was actually present during launch. Before describing that event though, I must mention the extensive preparations that preceded launch.

A number of lectures on the scientific mission to be carried by UK-6 were given by the scientists who had designed the experiments, and these were always of great interest to me as a keen astronomer. The main experiment was a cosmic ray detector from the University of Bristol. These were to be the first ever such measurements made in space. There were also two X-ray experiments on board – one from the University of Leicester, which could study variable X-ray sources with high time resolution, and the other, a joint venture between the Mullard Space Science Laboratory of University College London, and the University of Birmingham. These groups provided an instrument operating at low X-ray energies, viewing in space along the spin axis of the satellite.

A general knowledge of the experiments gives staff an appreciation of the work that goes into the production of the equipment on board the satellite. UK-6 was carrying on the X-ray investigations in space pioneered by UK-5, and the teams running both operations were essentially the same.

As well as lectures on the UK-6 science package, we were able to visit the Laboratories of Marconi Space and Defence Systems, the company which was building the satellite in Portsmouth. During that visit we all donned white cover-alls from head to foot. Such is the cleanliness required for all space missions.

Rehearsals

Before launch, several rehearsals were held, involving each team. Because of the possibility of unexpected delays, it was impossible to know which team would be on duty for launch, so each had to know the exact procedure to be followed, plus all contingency arrangements. Unknown to me, I was to be in the hot seat!

All command sequences for each experiment were constructed by computer. Each sequence would carry out one operation – such as operating a switch to change from one circuit to another. These sequences were sent, not only to Winkfield – our prime station – but also to other NASA stations that we might wish to use from time to time. During the launch phase, we would be following the satellite around the Earth for several orbits, switching from station to station, until we were sure that everything on board the satellite was OK.

Leaving out a few postponements, the night of the launch arrived, with the launch team now defined, including a colleague and me in command. For any launch, there is normally a complete turn-out of all the staff involved – even those who would be due in some twelve hours later. Such is the concern to see the satellite safely in orbit.

When all contingency plans were completed and the practising over, one had to wait and do the final checks. Computers were running, voice and data lines around the world were repeatedly checked, and at the launch site at Flight Centre, Wallops Island, Virginia, final preparations were made.

On the Control Centre desk lay the launch sequence, practised so many times, and a long list of suitable NASA ground station passes. We had scama phone links set up with NASA HQ, Winkfield, and there were groups elsewhere listening in, anxious to hear the progress. Simulated data was flowing from the ground station to monitor the quality of the data lines.

Countdown had proceeded, together with an increasing tension as the minutes to launch decreased. There was a hold shortly before launch – fortunately not serious – and the countdown continued. At thirty seconds to go, another hold! A minor problem, and after a few more seconds the counting continued – then launch!

Launch

During launch, both on-board tape recorders had been running in recording mode – so that the violence of launch would not de-spool the tapes. Following the successful placing of the satellite in orbit, it was renamed ARIEL-6. We followed it around the world, moving from station to station and monitoring the various systems. It was important to know that temperatures and voltages on board were as expected. The satellite had a battery which was charged during the sunlit part of its orbit, and which powered the experiments while the spacecraft was in eclipse. Such a vital housekeeping process is essential work on any satellite.

UK-6 had a successful operational life and considerable use was made of the other ground stations in the NASA network. Unfortunately, problems with the tape recorders did result in some lost data, but so much good data was collected during the lifetime of UK-6 that its controllers always considered it to be a challenging and rewarding satellite.

IRAS – the Infra-Red Astronomical Satellite

The IRAS project (Figure 1.3) was undoubtedly the highlight of my own personal involvement in satellite operations, but more importantly, was a milestone in the progress of astronomical knowledge. It was a major international collaborative effort whose results continue to be published as the backlog of years of results is processed. This section describes some of the events that occurred during the operational phase of IRAS, i.e. before it ran out of the liquid helium cryogen needed to cool the very sensitive detectors.

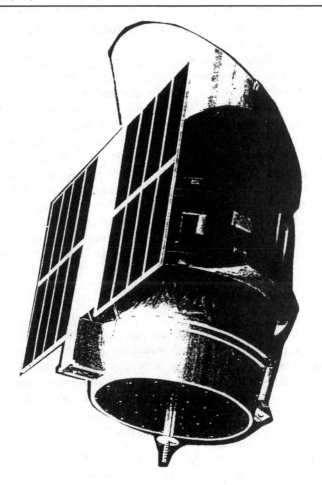

Figure 1.3   *The IRAS satellite*

Following the end of ARIEL-6 operations there was a period in which extensive preparations were carried out for the IRAS project. It was commanded from the Rutherford Appleton Laboratory, as had been the ARIEL satellites before it, and by essentially the same teams.

Work on the IRAS project began in 1974, with the National Aeronautics and Space Administration (NASA), the Netherlands Agency for Aerospace Programs, and our own Science and Engineering Research Council in consultation. The craft was launched on 26 January 1983 and remained operational for 300 days, running out of cryogen on 21 November the same year.

To put the project in perspective, glance at a diagram of the electromagnetic spectrum and you will see that most parts of it have been explored by one means or another. The universe has been studied by telescope, that is, in the visible part of the spectrum. In the drier parts of the world, high up in the mountains of Hawaii, there are telescopes that can see some little way into the infra-red; because water vapour in the atmosphere absorbs infra-red, you need to get the bulk of the atmosphere below you if you want to look at anything near those wavelengths.

When Galileo turned his telescope to the night sky he was using an optical aid to see the universe. His work was pioneering. Our IRAS team was the first to see the universe at infra-red wavelengths, so I was proud to be closely involved.

Dish for high frequencies

The experiments on board IRAS were provided by the Dutch and the USA; the UK provided the Control Centre and operating teams. The main antenna was a 12 metre steerable dish, installed especially for the project, and required because of the very high operating frequency required to transmit the satellite data.

The satellite would be collecting huge amounts of data on each orbit and this had to be stored for up to 12 hours before 'dumping'. A typical pass for IRAS would last about 15 minutes and, because of the preference to use only the middle few minutes to dump stored data, such data had to be transmitted at a very high rate. To effect this, the carrier signal had to have a high bandwidth, which one can calculate to be a few megahertz, so suitable frequencies for the carrier needed to be considerably higher. At such frequencies the most efficient signal collector is a dish.

Signal processing

On most scientific (or other) satellites, on-board equipment produces large amounts of data. This data has to be converted to a form suitable for transmission to the ground station. To do this, it must be efficiently coded (inserted) on to a radio wave called a carrier, and the more data there is to be coded, the higher the carrier frequency must be. The process of coding is called modulation. The satellite then transmits a modulated carrier wave to be received by any suitably equipped ground station.

The signal from IRAS, received by the dish, was first converted down to 50 MHz because, at this frequency, cable transmission losses are much reduced. In later chapters you will encounter the same principle in amateur work. Cables carried the 50 MHz signal across ground, into the Control Centre, then into the receivers. The original data was then to be de-modulated – the process of extracting data from the carrier signal. This initially processed signal was then fed into twin PDP-11/34 computers for further processing.

The dish had a relatively small beamwidth (the angle of reception in which it could receive data), so it had to be driven accurately. A tracking (computer) program was run during each pass, and in fact our IRAS dish could be driven either manually (by joystick, would you believe!) or automatically by computer. Before launch everyone in the operating teams had an opportunity to practise using the joystick guiding system to track a satellite – great fun, but a very serious job. We were determined to be ready for any eventuality.

Following a period of extensive software testing, simulated data processing runs, simulated faults, and other contingency operations, we were ready for launch and took our places on the night. Once again fate dealt an unexpected hand. Launch was delayed on a couple of occasions, putting my team on duty, with me once more as Satellite Controller. My colleague from UK-6 days, another shift leader, took the communications post next to me, so we were in total control, monitoring all the ground stations that would track IRAS from launch and during its first few hours of life.

On the night of launch I saw many well-known faces; the evening's schedule left us little time for socializing though, so after my team had taken over the operations we started preliminary checks of all equipment. There was a schedule of events for the launch, reproduced here.

Final preparations started officially at 30 minutes to lift-off, i.e. T-30:00. The voice line and two data lines were configured (set up).

- At T-25:00   Pre-test briefing completed – check with the ground station – they understand the content of the pass briefing messages.
- At T-15:00   Operations Supervisor confirms his transmitter is loaded with commands set up in a safety mode to avoid accidental transmission. He confirms the readiness of the 'hot back-up SCE' – a transmitter set with emergency command sequences. (SCE means spacecraft command encoder.)
- At T-14:00   Simulated (sim) spacecraft data is sent.
- At T-10:00   Sim data is removed and a test block of data from the SCE (transmitter) is sent.
- At T-02:00   Transmitter carrier switched on. Pulses rise with the adrenaline level. A final glance around the Control Centre to check that everyone is in place.
- T0   IRAS is launched.
- At T + 00:48   Launch ground station receives the first telemetry (signals) from IRAS:
  AOS announced. Within a few further seconds, data comes into the Control Centre; I announce 'Rutherford has AOS' (buzz of excitement heard close by).
- At T + 01:43   Go for command. First command loaded into the computer and transmitted to IRAS. The satellite responds

correctly – command is verified. I announce that to the Control Centre Manager, who announces over the local intercom 'The first command has been transmitted to IRAS. It has been verified.' A round of applause is heard as the tension of launch is released and there is much cheering.

- At T + 03:40 Following a series of commands, all verified, we have MECO (main engine cut-off). Stages 1 and 2 separate on schedule.
- At T + 04:00 Stage 2 ignition, then fairing jettison (outer parts of rocket).
- At T + 04:15 The pyro valve sequence is started. I transmit the enable command, then the fire command; both are verified, much relief.
- At T + 07:28 We have loss of signal (LOS).

Launch followed the scheduled sequence and after confirming that all was well, we waited for IRAS to come around to Chilton to give us our first pass. Launch tracking, and early commanding, were done using NASA stations.

Chilton Pass 1 was our first go at a complete operation. This did not involve the NASA network – it was completely up to us. The long-practised session started with the computers being thoroughly checked out, and the dish electronics re-tested. The report 'Chilton configured for Pass 1' was given to the Mission Operations Manager.

The first pass

At the expected time (AOS) we had been ready to go to manual tracking with the joystick, should the satellite not have been in the beam of the dish. IRAS did arrive! In the presence of each experimental team, on-board systems were checked out, section by section, without problems. The first command sequence was sent to start setting up the individual experiments. Although they were covered by a protective shroud and therefore not making any measurements, their safety was of paramount importance.

Solar panels were deployed – that is, extended outwards from the satellite, to allow them to sunbathe! The tape recorders were activated, and other checks made. In case commanding and checks were to take longer than expected, we had the additional ground stations at Alaska and Hawaii available. When we lost the IRAS signal, it would be picked up by the next station along the route. IRAS and the teams worked well together and there were no problems.

The next significant event was to 'blow-off' the protective cover. Yet another quirk of fate; proceedings and checks went so well in the post-launch phase that the cover ejection date was brought forward by one day – when I was on duty!

The pass went without a hitch and the all-important command to blow off the cover was sent. Immediately the temperature inside the craft dropped like a stone, as anticipated, and a delighted team of astronomers became the first people to ever see the universe in the infra-red.

2 Select your satellites

This chapter provides a look at different types of satellites to identify those which can be monitored or decoded, frequency bands and the required equipment.

Having read of professional satellite operations, you could feel isolated from this world of high technology, perhaps believing that you could never have any part in it. You would be wrong. Although it is unlikely that you could just step into a job in space research, there is nothing other than acute financial hardship to stop you involving yourself in this field, at an amateur level.

The purpose behind my describing satellite operations in the previous chapter was to explain the manner in which satellites are controlled and the data processed. You should find the following chapters of help should you now wish to participate in satellite operations in your own home, as I have for some years.

This chapter helps to identify the most suitable satellites for your interests, by describing the various groups and explaining what type of equipment is required to receive them. Depending on the type of satellite, and your budget, you may then go on to process the data.

You already have an idea of the general facilities needed to do some of the work previously described. A computer is almost essential, and a satellite predictions program can show you where the satellites are. Suitable receivers – and, of course, an antenna or two – are the next stage. You don't *have* to analyse the data – you can collect it for future use, or simply keep records of which satellites are operating. Some of this information is currently available from a variety of sources and published in specialist magazines – all will be revealed.

So what satellites can we receive? A large number transmit telemetry – data – or other signals. The real questions are – which satellites are of interest for possible data analysis, which are worth listening to, and which can be ruled out because they require the use of equipment that is either too expensive or unobtainable?

Satellite monitoring is an open-ended project. Many, like me, devote several hours each week to it, and others spend just a few. Your first entry to the field might be simply to obtain a shareware/public domain computer program, and 'get the feel' of satellite tracking by running it on your PC. The cost of this may be little more than that of a stamp!

Money can be spent on equipment over a period of time, as finances permit; you do not have to buy everything at once and risk an expensive mistake. Have that holiday and do some reading – and planning!

Which are the best satellites?

Although there are several thousand artificial satellites in orbit, only a few hundred or so are actually in operation at any one time. They can be classified in various ways – the nature of their data (scientific, reconnaissance, entertainment etc.) or by their transmitting frequencies (20 MHz to 20000 MHz). Reception of satellites in each group can be costed and analysed on the basis of how interesting or useful their data is. This perception varies from person to person.

Someone once commented that he had spent hundreds of pounds on a weather satellite receiving station, and that evening saw the same pictures on the news. He wondered whether he had done the right thing!

Satellite groups

The following types of satellite can be considered. It is by no means exhaustive but it helps to put things into perspective.

- Amateur radio satellites
- Amateur science satellites
- Direct Broadcast (Television) Satellites (DBS)
- Domestic and international communications satellites
- Manned American and Russian satellites
- Navigation satellites
- Oceanographic research satellites
- Reconnaissance satellites
- Scientific satellites
- Solar system satellites – space probes
- Weather satellites

Each group can be further sub-divided; navigation satellites are closely associated with certain military satellites, e.g. the CIS COSMOS series, which uses the 150 MHz band.

This list could also be arranged in terms of frequencies, but many of the satellites listed above transmit on more than one frequency.

Amateur radio satellites

There are a large number of satellites that have been funded by the voluntary radio amateurs' organization AMSAT, of which the British section is called AMSAT-UK. For details about contacting this organization please see the Appendix.

The Orbital Satellite Carrying Amateur Radio (OSCAR) series of satellites and the then Soviet equivalent, Radio Sputnik (RS) satellites are all available for general monitoring, due to the work done by amateur radio organizations around the world. These satellites sometimes carry equipment designed for educational purposes, and we shall look at UoSAT-2, an excellent example of this type.

For monitoring purposes, many of the frequencies used by the OSCAR satellites can be received with scanners, when suitable antennae are used. Those amateurs who are licensed to transmit, and who have suitable transmitting and receiving equipment, can use the OSCAR satellites for communication. For further details, again contact AMSAT-UK (see Appendix 1).

Downlink frequencies

The following are some of the frequencies available for monitoring amateur radio satellites. Lists are freely available from many sources including, of course, AMSAT-UK. Uplink frequencies are not included because this list is to aid monitoring only.

OSCAR 11/UoSAT 2
Circular, polar orbit.
Telemetry and other data on 145.825 MHz. Left-circular polarization. The astronomical and engineering data transmitted by UoSAT-2 makes this invaluable for monitoring.

RS-10/11
Beacons: 29.357, 29.403, 29.407, 29.453, 145.857, 145.903, 145.907 and 145.953 MHz.

AMSAT-OSCAR 13
High-altitude elliptical orbit.
Beacons include: 145.812, 145.985, 435.652 and 2400.664 MHz
Some of the telemetry contains interesting engineering data. Further details about its format and content can be obtained from AMSAT-UK. The main operations with this satellite are communications between amateurs and so the downlink frequencies can be monitored in the 145, 435 and 2400 MHz bands.

WEBERSAT-OSCAR 18

Low altitude, circular, sun-synchronous, near-polar orbit.

Primary objective is to take, store and transmit Earth images using an on-board CCD system, in a format using amateur satellite packet radio techniques.

Downlink frequencies: 437.102 and 437.075 MHz.

LUSAT-OSCAR 19

Low-altitude, circular, sun-synchronous, near-polar orbit.

Primary objective is to provide the world-wide amateur radio community with a low-Earth orbit, satellite store-and-forward packet radio message system.

Beacon 437.127 MHz; Downlink frequencies: 437.153 and 437.125 MHz.

FUJI-OSCAR 20 (FO-20)

Low-altitude, elliptical, non sun-synchronous orbit.

Packet radio store-and-forward communications with other transponders.

Beacon: 435.795 MHz.

AO-21/RS 14

Multi-function satellite.

Beacon: 145.948 MHz (CW), 145.838 and 145.80 MHz digital.

SARA-OSCAR 23

Low-altitude, circular orbit.

A French amateur radio astronomy satellite monitoring radio emissions from the planet Jupiter in the 2–15 MHz band.

Downlink: 145.995 MHz.

Amateur science satellites

The most important satellites, from an amateur scientist's point of view, are probably the University of Surrey satellites (UoSATs), of which there are several. UoSATs are excellent for monitoring the space environment, in much the same way as with UK-5 and UK-6.

UoSATs carry scientific experiments, so a meaningful research programme can be carried out. This is where the amateur scientist can actually become involved in research at the highest level. If you write your own software for data analysis, using the vast quantities of data broadcast by these satellites, the experience gained is second to none.

UoSAT-2 provides a variety of telemetry formats using the same frequency; real-time data, data recorded during a whole orbit, and satellite Kepler elements (see Appendix 2) are all included sequentially.

As equipment and setting-up progress, you will be able to accumulate meaningful results. You may then decide to invest more money to build an advanced set-up capable of decoding other telemetry. Study this project and then check out project costings for yourself.

If your interests include space physics, then to whet your appetite, Figures 2.1–2.3 show three computer print-outs using data received from UoSAT-2 in early 1988. They show near-Earth magnetic field variations measured along three axes by the satellite.

Direct Broadcast (Television) satellites

In the early 1970s I was engaged in research on solar physics, using radiometers. These were telescopes operating at very high frequencies – from about 1 GHz (gigahertz – one thousand, million hertz) right up to 80 GHz.

The different radiometers were located at the Solar Physics Observatory at the Radio and Space Research Station at Datchet. These allowed us to monitor bursts of solar radio energy over a wide range of frequencies. By its very nature, the same equipment was also collecting data on the way in which the atmosphere absorbs these high frequencies.

Colleagues used the same charts to analyse this data, in order to identify

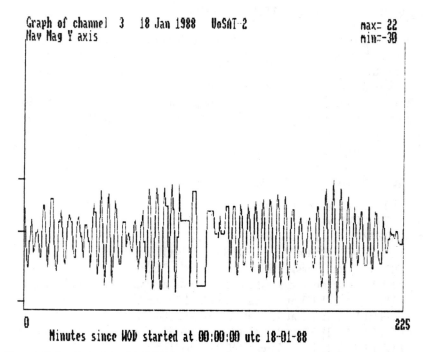

Figure 2.1 *Data from UoSAT-2*

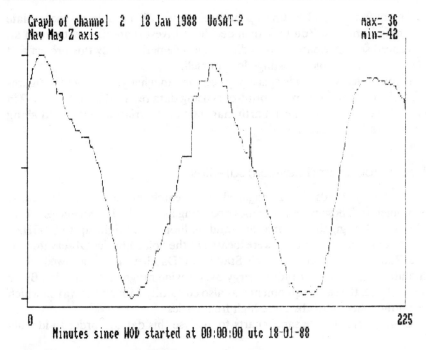

Graph of channel 2   18 Jan 1988 UoSAT-2   max= 36
Nav Mag Z axis   min=-42

0   225
       Minutes since WOD started at 00:00:00 utc 18-01-88

Figure 2.2   *Data from UoSAT-2*

sections of the frequency bands to be used 'in the future' for satellite-to-ground television broadcasts. That future arrived a few years ago when the World Administrative Radio Conference (WARC) divided this new section of the radio spectrum between participating nations and utilities.

The following frequencies are among many now used for television transmissions. Their signals are sometimes scrambled to restrict viewing to countries where the channels are legally allocated. These various types of encryption are fascinating to investigate, but beyond the scope of this book.

Frequencies (approximate)

- INTELSAT 602 at 63° east: 10.975–11.173 GHz
- INTELSAT 604 at 60° east: 10.9–11.8 GHz
- GORIZONT 17 at 53° east: 11.525 GHz
- TURKSAT 1B at 42° east: 10.980–11.144 GHz
- DFS 1 (Kopernicus) at 23.5° east: 11.5–12.7 GHz
- ASTRA (four satellites – with more to come) at 19.2° east: 10.714–11.7 GHz
- EUTELSAT II F3 at 16° east: 10.972–12.54 GHz
- EUTELSAT II F1 at 13° east: 10.972–12.584 GHz
- EUTELSAT II F2 at 10° east: 10.9–11.6 GHz
- EUTELSAT II F4 at 7° east: 10.9–11.6 GHz
- INTELSAT VA F12 at 1° west: 11.0–11.7 GHz

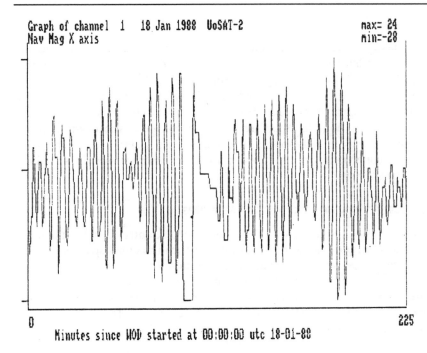

Graph of channel 1   18 Jan 1988   UoSAT-2   max= 24
Nav Mag X axis   min=-28

0   225
       Minutes since HOD started at 00:00:00 utc 18-01-88

Figure 2.3 *Data from UoSAT-2*

- TELECOM 1C at 5° west: 12.5–12.7 GHz.
- OLYMPUS 1A at 18.8° west: 12.1 GHz
- INTELSAT VIF4 at 27.5° west: 10.9–11.1 GHz

New launches occur from time to time. In each case, the frequencies used have a certain characteristic, called the polarization. This refers to the direction in which the radio signal is vibrating – whether left or right – so searching for these satellites has to be done in a methodical manner. I do an occasional frequency scan with my own ASTRA receiver, to see if I can find test broadcasts from new channels. Various monthly magazines, such as *Short Wave* and *What Satellite*, include different aspects of satellite television, and lists of frequencies and satellites are published.

Recent years have seen a rapid expansion in the number of both domestic and commercial users of satellite television. How intriguing to see the outcome of that early research done in Datchet.

One can now walk into many large city stores and purchase a complete satellite television receiver with dish, so this should be included amongst the various satellite groups that can be monitored.

Compared with possible spending on other satellite monitoring, television satellites are now economical. For less than £200 (the cost varies widely), one can purchase an ASTRA receiver, with dish antenna, low-noise amplifier, and get perfect pictures without any major problems.

Domestic and international communications satellites

Books have been devoted to communications satellites, and references to these are given in a later chapter. Different rules apply in different countries for the reception, by 'non-authorized people', of such satellites.

In the USA the relevant frequencies are freely published and such lists are also freely available in the UK; however, caution and common sense must be exercised. With suitable equipment one can monitor transmissions from communication satellites – though, with a huge range of frequencies, and data encryption often used, one has to accept that there are better ways of spending hard-earned cash!

Manned American and Russian satellites

Back in the late 1970s and early 1980s, our Satellite Control Centre often took the opportunity to 'patch in' to various American manned space missions. Conversations between ground mission operations centre in the USA and the shuttle flights were most interesting to monitor.

Some excellent articles have been written about monitoring both the American Shuttle and the Russian manned space station MIR and both can be heard with fairly conventional scanning receivers.

USA Shuttle

There are various ways to monitor shuttle activities. Perhaps the simplest is to use a domestic receiver to listen to 'Voice of America', the radio station which, amongst other items, provides reasonable coverage of the American space programme. It broadcasts on a number of published frequencies including 6.024 MHz. Although I have monitored several of the frequencies given here, conditions and project requirements vary, so nothing can be guaranteed.

Shuttle flights are generally of fairly short duration, and when the inclination of the orbit is under about 50°, the craft will not pass over Britain. On such occasions, we can still monitor communications by tuning into the re-broadcasting frequencies used by the Goddard Space Flight Centre Amateur Radio Club (GARC) which transmits NASA Select audio (see later), using amateur radio.

These frequencies are unofficial but may serve as a guide during shuttle flights. Data comes from NASA Spacelink, and is provided by the NASA Educational Affairs Division:

3.86, 7.185, 14.295, 21.395, 28.650 MHz, All USB (Upper Side Band)

For the direct monitoring of shuttle transmissions (tuning in to the shuttle when it passes over the UK) the following frequencies, provided by NASA publications, may be heard:

- 259.7 MHz air to ground or suit to orbiter
- 296.8 MHz air to ground or orbiter to suit
- 279.0 MHz suit to orbiter or suit to suit

- 243.0 MHz standard military aircraft emergency frequency
- 2205.0, 2217.5, 2250.0, 2287.5 MHz frequencies also used

'Suit' refers to extra-vehicular activities, in which transmissions are made while outside the main spacecraft. It will be appreciated that a suitable antenna is needed for receiving frequencies in different bands. For more information on the shuttle, the relevant address is given in Appendix 1 near the end of this book.

Shuttle audio retransmissions

Shuttle audio is re-transmitted by the following amateur radio stations:

Station	Centre	VHF	10 m	15 m	20 m	40 m	80 m	
WA3NAN	GSFC	147.450	28.650	21.395	14.295	7.185	3.860	
W6VIO	JPL	224.040		21.280	14.282	7.165		
K6MF	ARC	145.585				7.165	3.840	
W5RRR	JSC	146.640	28.495	21.350	14.280	7.227	3.850	
AK8Y	LERC	145.670	or 147.195 (alternate)					
W1AW	ARRL	147.555	28.0675	21.0675 18.0975	14.0475	7.0475	3.5815	1.818
KA9SZX		146.880 (video at 426.250)						
K4GCC		146.940						
WA4VME		145.170						

All frequencies are in MHz. Use FM on VHF, USB on 10–20 m, LSB on 40–80 m.
WA3NAN – NASA Goddard Space Flight Center (GSFC), Greenbelt, MD.
W6VIO – NASA Jet Propulsion Laboratory (JPL), Pasadena, CA.
K6MF – NASA Ames Research Center (ARC), Moffett Field, CA.
W5RRR – NASA Johnson Space Center (JSC), Houston, TX.

Monitoring MIR

It has been quite easy to monitor the manned space station MIR (operated by the Commonwealth of Independent States) for a number of years, using standard scanners. The voice link used for routine conversations between MIR and the ground station is un-encrypted FM, transmitted on 143.625 MHz. This is within the range of countless radios, being only just outside the amateur radio band.

As a result of considerable publicity within the amateur radio community, many people, including me, listen regularly to these conversations, and on at least two occasions, stressed voices could be recognized, even if the Russian language was not comprehended.

In recent years, various frequencies have been logged from the different vehicles associated with the Russian manned space programme, so they are listed here:

- 143.625 MHz Voice link with MIR
- 121.75 MHz Soyuz
- 166.12 MHz Progress
- 145.50 MHz Amateur band transmissions from MIR

Navigation satellites

There have been articles published by various researchers studying the Russian navigation satellite system and, in this section, reference must be made to these satellites because they can be easily monitored.

If you have a scanner which covers the 150.0 MHz band, then searching for signals between about 149.8 and 151.0 MHz will yield a rich reward during just a few hours' scanning. The following are those spot frequencies currently being used by the listed COSMOS satellites:

- 149.91 MHz COSMOS 2184
- 149.94 MHz COSMOS 2218, 2279
- 149.97 MHz COSMOS 2239, 2266
- 150.00 MHz COSMOS 2230
- 150.03 MHz COSMOS 2233

In addition:

- 399.96 MHz COSMOS 2184
- 399.84 MHz COSMOS 2279, 2266, 2239, 2218
- 399.92 MHz COSMOS 2195
- 400.00 MHz COSMOS 2230
- 400.08 MHz COSMOS 2233

These satellites form part of the military and civil navigation system, and can be heard regularly. The COSMOS satellites listed above may well change during future months and years. Each has a replacement which may be operated from time to time. I have spent a considerable time monitoring this group simply out of interest, obtaining Kepler elements from a number of sources and logging satellite change-overs.

Each satellite appears to use both frequencies simultaneously, and this can aid in positive identification. The telemetry from the satellite can be decoded – a fascinating project – but beyond the scope of this book.

Oceanographic research satellites

Not long after I had bought my first good quality scanner, purpose-designed for weather satellite reception, I heard and decoded a picture from a satellite transmitting on 137.40 MHz. It turned out to be one of the COSMOS series, number 1766. Investigation and discussion with a friend identified the satellite, and I later discovered other COSMOS satellites which are part of the Russian Oceanographic satellite system.

Much has been written elsewhere about the equipment carried by these, following a reduction of the intense secrecy that originally surrounded the operations of Russian satellites. We shall look at these satellites in more detail in a later chapter.

Reconnaissance satellites

Various articles have been written about early work involved in identifying these satellites, and a glance at the recommended reading section (at the end of Chapter 4) will provide details. Information has been obtained from a variety of sources, including Geoffrey Falworth, on the different satellites in this series. They can be monitored on a number of frequencies:

- 19.195, 19.994, 20.005 MHz

The art of monitoring these satellites deserves a book (or at least a chapter) to itself, and is outside the scope of this book. As on other occasions, these frequencies can be found on most good quality HF receivers.

Scientific satellites

Many satellites have carried scientific equipment into orbit, in order to get above the Earth's atmosphere and radiation belts. Satellites such as the Hubble Space Telescope (HST) and the International Ultra-violet Explorer (IUE) are just two examples.

In general, telemetry containing the scientific data is transmitted at very high frequencies. In the previous chapter on IRAS, I mentioned the process of modulating the high-frequency carrier with data. This data includes house-keeping information – engineering data, such as battery voltages, as well as scientific measurements, such as photo-electric counter data. It follows that amateur equipment will be unable to decode this, although on some occasions it is possible to monitor the beacon. This happened with UK-6, which I heard in 1986 using a scanner. It was an old friend, though I did not realize its identity for several months. I was able to monitor the 137.56 MHz beacon until shortly before the satellite re-entered the atmosphere.

I left this frequency selected in my scanner, and to my utter amazement, I heard the same signal again during 1991! Following a suggestion from Geoffrey Falworth, enquiries for Kepler elements for earlier satellites eventually enabled me to identify the satellite as X3, or PROSPERO, an early British satellite, long since unused. This illustrates how the unexpected can still be received in this band.

Weather satellites

This book describes the equipment needed to set up a comprehensive weather satellite receiving station, and should include all the information required for the project.

The polar orbiting satellites, including the oceanographic ones, use the 137–138 MHz band for picture transmissions. The 1691 MHz band is used by NOAA satellites for high resolution pictures. Geostationary weather satellites use the 1691 MHz band for both low resolution and digital, high-resolution, pictures.

Solar system satellites – space probes

A number of countries have launched space probes, some orbiting the Sun, and therefore satellites of it (such as the Earth), others – such as Ulysses – making complex trajectories to take measurements from special positions within the solar system.

All of these probes carry transmitting hardware to enable their measurements to be radioed back to Earth. The frequency bands are very high and the power of the signal is extremely low; the result is that amateurs are not able to receive this telemetry because of the extremely high cost of the hardware required to set up a suitable receiving system. Space probes normally have a dedicated ground station set up to receive and process the telemetry, as previously described in Chapter 1 on UK-5, UK-6 and IRAS satellites.

My own path

The route that I took into satellite monitoring, in my amateur capacity just two years after leaving professional satellite work, started with the UoSATs. My enquiries indicated that simple receivers for these satellites could help me achieve several aims fairly quickly. For £150 I bought a receiver with antenna, which included a decoder to provide suitable data for a computer. A few hours after receiving the equipment, I heard my first satellite since IRAS – I was elated!

I enlisted the help of Timothy, my son, to assist with writing programs for the BBC computer. I had been writing software for mainframes since 1968, but not for the BBC micro. Within a short time we had produced graphs of the magnetic field of the near-space environment as measured by sensors on UoSAT 2.

A few weeks later I bought a receiver, or rather the bits, for reception of the weather satellites, and quickly found myself able to hear the NOAAs and METEORs. The hardware decoder was a major headache but help from the local radio club soon had that working; pictures were received every day. Building a home-made dish and buying a down-convertor allowed me to hear METEOSAT. Then I bought a computer and commercial decoder. And then ...

3 Weather satellites

In this chapter we have a brief look at the history of weather satellites, look at the NOAA polar orbiters, their transmissions and their telemetry format. EUMETSAT's plans for polar orbiting satellites are given.

The beginning

The first artificial satellite was Russia's Sputnik 1; its launch effectively started the 'space age' on 4 October 1957. It weighed about 84 kg, a sphere of 580 mm diameter. I was then 12 years old, and remember announcements on the television news that day, which perhaps triggered my fascination with space.

The Americans were working on their own satellite launching programme, which had begun at the end of World War 2. On 1 April 1960 they launched TIROS-1 (Television and Infra-Red Observation Satellite) – their first craft designed specifically for monitoring weather. TIROS-1 was very basic by today's standards, but hardware developments led to the launch of ITOS (Improved TIROS Operational Satellite) on 23 January 1970.

The Russians also entered the field of meteorological monitoring, and other nations including India and China, have since launched weather satellites. This chapter looks at the American scene.

Early American weather satellites

The 'Automatic Picture Transmission' system (a.p.t. – to be discussed shortly) is used on all weather satellites, but it was first tested in 1963, on TIROS-8. The following year, NIMBUS-1 was launched on 28 August, followed by ESSA-2 on 28 February 1966 (ESSA was the Environmental Science Service Administration). The original vidicon cameras of the earlier satellites, were replaced by radiometers on NOAA-2, launched in 1972.

Earlier ITOS satellites were in almost circular, sun-synchronous orbits (see later), some 1450 km high. Following TIROS-10, the first of the following series of TIROS satellites was called TIROS-N, launched on 13 October 1978 to an 850 km high orbit. Data was then collected along a swath width (the distance across the width of the view below the satellite) of about 2700 km, so global coverage was achieved twice each day.

Satellite names

There can be some confusion about satellite names, due to the practice of having different pre-launch and post-launch names. Successfully launched satellites are numbered consecutively, so unsuccessful launches do not have a corresponding post-launch name. Until 1963, satellites were designated by year, with the addition of a Greek letter. Vanguard 2 is listed (in NASA's Satellite Situation Report for 1992) as 1959 Alpha 1. Vanguard 3 (the next successful launch) is 1959 Eta 1.

From 1963, there was a change in the convention used for naming satellites. The year is retained, but the object itself is designated according to its part – so the satellite payload is called A, the rocket body, if identified, is called B, and debris called C, etc. TIROS-7 was launched on 27 June 1963, taking the International Designation 1963-024A. NASA also allocates a catalogue number to each identifiable part – so TIROS-7 has the number 604.

TIROS-E was the first of the advanced TIROS satellites (AT-N) and, after launch on 28 March 1983, was renamed NOAA-8. TIROS-F became NOAA-9, following a successful launch on 12 December 1984. Still in operation, at autumn 1994, NOAA-9 was re-classified (in 1992) as an experimental satellite but has stopped transmitting a.p.t.

These later weather satellites carried new equipment, which included the Advanced Very High Resolution Radiometer (AVHRR) instead of the VHRR used on previous NOAA craft. The AVHRR features 'in-house' (on-board) data processing. After scanning the earth below, in five channels (instead of just three used as in the VHRR), data are digitally processed with both on-board and external calibration targets to provide reference temperatures. The NOAA satellites carry NiCad batteries (nickel and cadmium) to supplement the solar panels, by providing power during spacecraft night.

NOAA–GOES: two satellite systems

The National Oceanic and Atmospheric Administration (NOAA) assumed the responsibilities for weather forecasting, the issue of warnings, and space environment monitoring (such as solar flares). To perform these tasks, two series of satellites are used, the polar orbiting ones (currently NOAAs-9, 10, 12 and 14) and a series of geostationary satellites, called GOES (Geostationary Operational Environmental Satellite).

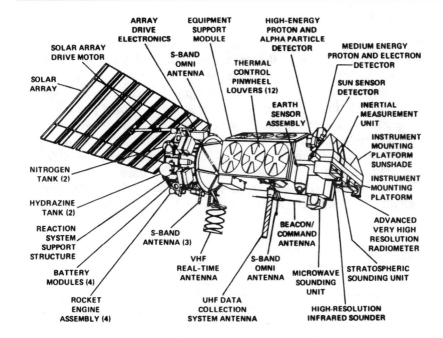

Figure 3.1 *NOAA weather satellite*

The two groups are complementary, the first providing world-wide coverage at high resolution, over a large spectrum (visible and thermal imagery), and the geostationary group having the additional ability to continuously monitor storm areas, particularly near the tropics.

Polar-orbiting satellites–the American NOAAs

The advantage of a 'polar' orbit is that the satellite covers almost the whole Earth each 24 hours, due to the rotation of the planet beneath. By adjusting the height and inclination of the orbit, periods of sunlight experienced by the satellite can be balanced with the power requirements of the on-board systems.

By noting the orbital characteristics of a satellite group, one can sometimes identify the mission requirements, at least to some extent. Studying the orbits of the NOAAs reveals that they pass over every place on Earth, at about the same local time each day–they are 'Sun-synchronous'. Use a satellite predictions program (see Appendix 3) to list the pass time for any NOAA satellite, e.g. NOAA-14. You will find that passes are northbound during the afternoon (between about 1200 and 1500 local time), and are southbound during the morning.

Sun-synchronous orbits

Analysis of pass times for NOAA-12 (and the other NOAA satellites), during a period of a few days, shows that each pass occurs a few minutes earlier on successive days, and has a slightly different maximum elevation. Although the times of individual passes change, the group of passes as a whole remains within the given times. In other words, the satellites' orbital planes precess, keeping pace with the Sun. This type of orbit is called Sun-synchronous.

Longer-term monitoring of these orbits reveals that an occasional small correction to the orbit (made using on-board thrusters) is required, or the orbits slowly drift, gradually becoming non-synchronous.

The times shown in Table 3.1 are those times at which the satellite passes at maximum elevation (overhead). On any given day the satellite will probably pass earlier or later.

Table 3.1   *Pass characteristics*

Satellite	Time and direction	
NOAA-9*	Southbound	Northbound
	1000 UT	2030 UT
NOAA-10*	0700 UT	1730 UT
NOAA-11*	0600 UT	1650 UT
NOAA-12	0810 UT	1850 UT
NOAA-13*	0300 UT	1300 UT
NOAA-14	0300 UT	1300 UT

*Satellite not now transmitting a.p.t.

This data originally related to November 1994, with additions during December, following the successful launch of NOAA-14. NOAA-10 was powered off in early December, then unexpectedly switched back on. NOAA-11 has since been powered off, except (as at March 1995) its beacon! Activity may continue to change over a period of months.

As seen from the table of pass times, all the NOAAs are in non-identical, but still Sun-synchronous, orbits. A study of the pass 'windows' (periods of time during which the satellite can be above the horizon) shows that NOAAs 10 and 12, which both transmit a.p.t. (normal resolution pictures) on the same frequency (137.50 MHz), will occasionally overlap; in other words they can both be above the horizon at the same time. When this occurs, priority has been given to NOAA 12, so the a.p.t. from NOAA 10 is normally switched off. On some occasions, for reasons which are not obvious, NOAA 10 a.p.t. has been left transmitting, causing mutual interference.

Advanced very high-resolution radiometer (AVHRR)

A brief mention of transmissions on 1691 MHz was made in the previous chapter. Frequencies in this band are used by the polar-orbiting NOAA satellites to transmit picture data of high resolution – down to 1.1 km. The data used for the VHF signal known as automatic picture transmission (a.p.t.) is derived entirely from the AVHRR system.

The AVHRR systems on current NOAA polar satellites consist of a 203 mm Cassegrain telescope, receiving radiation from a beryllium scanning mirror, rotating at 360 r.p.m. Incoming radiation – whether from Earth, cold space, or a reference temperature within the satellite – is split into five beams, falling on five separate sensors, sensitive to the bands indicated. Part of the mirror's spin provides a reference level, when the telescope is looking into deep (cold) space.

The five channels include visible light, reflected infra-red, and three thermal channels. This multi-channel data is digitized by the on-board systems processor and the resulting processed data is transmitted on one of a number of frequencies: 1698.0, 1707 or 1702.5 MHz, depending on the satellite.

The channels monitored by the AVHRR scanner have been carefully chosen to look at the Earth through 'windows' – those gaps in the spectrum through which radiation penetrates the Earth's atmosphere with minimal absorption.

AVHRR Wavelengths

Our atmosphere forms a very protective layer – and thank goodness it does! Without this protection life on Earth would be either impossible or, at best, extremely hazardous. The atmosphere also protects us from showers of meteors, and from many dangerous forms of radiation. In order to study these radiations we have to send instruments above the atmosphere.

Table 3.2 *Channel wavelength (micrometres), main uses*

Channel number	Wavelength (μm)	Radiation type	Main uses
1	0.58–0.68	vis	Weather forecasting, cloud, snow and ice monitoring
2	0.725–1.10	vis-nir	Location of water, ice and snow melts, vegetation and agriculture monitoring
3	3.55–3.93	ir	Sea surface temperatures, night clouds, volcanic activity, fire monitoring
4	10.30–11.30	ir	Sea surface temperature, soil moisture
5	11.50–12.50	ir	as above

The atmosphere does not, of course, absorb *all* radiation. Gases in the atmosphere absorb certain frequencies, so it is possible to identify those frequencies not so strongly absorbed. These bands of frequencies are called 'windows' because they allow radiation to come through the atmosphere.

By careful selection of these windows, AVHRR equipment is able to provide measurements of temperature with high accuracy. Additionally, the data can be used to assess agricultural and vegetation situations because water measurements from the data can be analysed and its presence or absence recorded.

Visible

The bands are carefully chosen because of their properties. Visible-light images show clouds, land and ocean by their reflected light; this in turn depends on albedo – that percentage of light which the object reflects, compared with the original (incident light). Clouds reflect between 50 and 65% of the light they receive; snow reflects between 55 to 90% (depending on age), and the oceans reflect a mere 2–7%, depending on the state of the sea.

Infra-red

The original scanner signal is processed so that it produces a large signal (maximum amplitude) where the temperature is the highest; this results in cold cloud tops appearing as white, giving clouds a similar appearance in both visible and infra-red images. Seasonal changes in Mediterranean countries can be monitored most effectively in infra-red; we see the oceans warmer (darker) than the land in winter, but cooler (lighter) than land in summer.

Water vapour

This band operates where water vapour radiates strongly. The radiation intensity depends on the amount of moisture in the atmosphere, so a dark area indicates that the troposphere is dry.

Automatic picture transmission (a.p.t.)

As briefly mentioned, this transmission system dates back to the 1960s, and because of its successful adoption by other countries involved in weather satellite operations, it is possible to decode all such (a.p.t.) pictures, using similar equipment.

Sensors on board the satellite detect radiation from the Earth, converting it to an analogue signal. This signal varies in amplitude depending on the intensity of the radiation detected by the sensor. This varying signal amplitude modulates a 2.4 kHz carrier, properly called a sub-carrier. This sub-carrier, now containing picture information, is then used to frequency modulate the main carrier in the 137 MHz band, e.g. 137.50 MHz (for NOAA-12). The maximum deviation, within the signal modulation, is for peak white, and corresponds to 17 kHz. The minimum deviation is not zero – which would cause problems for receivers unable to lock on to a vanished sub-carrier – it is about 5% of the peak white level.

NOAA weather satellites have used consistently accurate 2.4 kHz sub-carriers, resulting in well-synchronized pictures. The METEORs have not been so meticulous. Series 2 METEORs used sub-carriers which fluctuated by up to 10%; the Series 3 METEORs visible-light pictures have been more stable. METEOR modulation levels have also varied from the ideal described above – see Chapter 4.

The signal transmitted at VHF frequencies, in the a.p.t. format from NOAA satellites, comes from the original AVHRR scanners which produce high-resolution images. The process of a.p.t. production involves using the h.r.p.t. signal and degrading it electronically, reducing the data content to a form suitable for transmission at VHF frequencies. Degrading involves removing every third line, so in this form (a.p.t.) the resolution is reduced to about four kilometres.

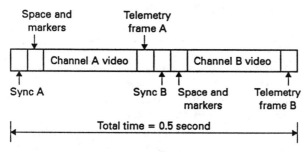

Figure 3.2 *NOAA a.p.t. format*

The a.p.t. transmission from NOAA satellites takes the form of a picture containing two sections, each of which is further sub-divided. Listening to the signal one can identify these two parts – a 'clip' followed by a 'clop'. The whole data line (illustrated) takes half-a-second to complete, and during that time several individual components are transmitted.

Starting from the left side of the diagram, the first tone is actually a synchronization pulse, called sync A (signifying that this precedes channel A). This allows software (running on a computer) or a framestore, to recognize the following section (i.e. to know what comes next). The channel A sync pulse consists of seven cycles of a 1040 Hz tone.

The next section contains the minute markers, separated by space. Minute markers actually consist of four lines, two white and two black – repeated each minute. An experienced ear can recognize them quite easily. The actual video data is contained in the next section; this is the sensor data which is extracted from the original AVHRR sensors and degraded to enable its transmission at VHF frequencies. This section takes up most of the width of this half of the picture.

The next section (called telemetry frame A) is a part of a wedge which contains a calibrated level that can represent a temperature or grey level – see later. Software can use this information to make qualitative measurements in the video picture.

That completes the first quarter-second, so the next section starts with a further tone burst at a different frequency – 7 pulses of 832 Hz. This precedes channel B video, and so is called the channel B sync. It is similarly followed by a section containing minute markers and space. They are followed by channel B video information and, finally, the telemetry wedge containing frame B calibration data.

As each line is received, the picture is slowly built up on a screen and you can see how the minute markers appear and how the telemetry wedges change as the calibration sequences change.

These wedges total 16 different levels – the picture then repeats, starting again from wedge one. The whole sequence repeats after 128 lines of half-second data – a total of 64 seconds.

Calibration data

As mentioned, the a.p.t. image contains a set of calibration levels – the sequence of shaded rectangles shown. Blocks 1 to 8 are shaded from darkest to lightest, the first five representing channels 1 to 5. Block 9 is black, representing zero modulation. Blocks 10 to 15 correspond to calibrated grey levels, and block 16 shows the channel in use for that image section. Match this grey level (block 16) to the corresponding block in the calibrated group (blocks 1 to 5), and this number gives the channel in use – see Fig. 3.3.

Compatibility with other a.p.t. satellites lies not only in the frame width time being either identical (half a second) or a multiple, but in the way that the original picture data is always amplitude modulated onto the same 2400 Hz sub-carrier which then frequency modulates the main VHF (radio frequency) carrier. In this way, METEOSAT, GOES, OKEAN and the METEORS all provide picture data using 'a.p.t.' format, in one form or another. The only real difference is in the actual content of the picture and its resolution. Details about the picture content and format of METEOR, OKEAN and the geostationary satellites are given later.

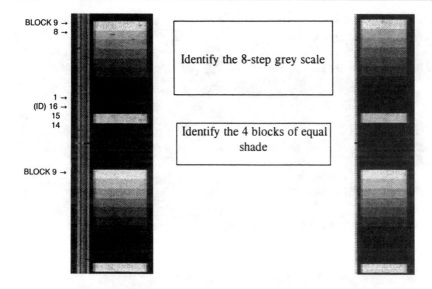

Figure 3.3 *NOAA a.p.t calibration scales*

Outgassing

Thermal sensors have to be cooled because their operation relies on a sensitivity to temperature; as a consequence of such low-temperature operation (105°K), condensates form on the scanner aperture. For purposes of maintenance, these surfaces are periodically allowed to warm up, a process which removes the impurities and surface contaminants, and is called outgassing. During this time the infra-red channels are inoperative. Consequently, at this time, each NOAA satellite changes from transmitting i.r. and visible, to transmitting the two 'visible-light' channels (1 and 2), side-by-side. The picture (Fig. 3.4) shows this operation.

Sounds incorporated

Because the 2400 Hz carrier of a.p.t. lies in the audio part of the spectrum, we can hear any changes in the content of the signal caused by various characteristic weather patterns. More on this in the next chapter.

Other equipment

Other instrumentation carried by the NOAA craft includes the *High Resolution Infra-Red Sounder* (*HIRS*) which measures the vertical profile of atmospheric temperature, and monitors the total ozone content and water

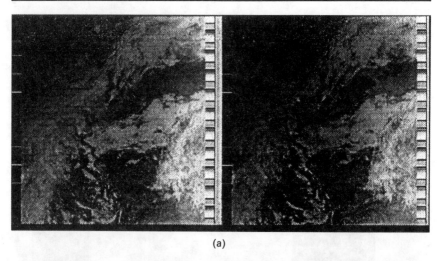

(a)

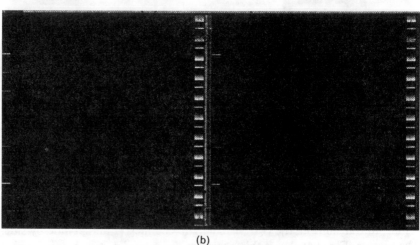

(b)

Figure 3.4 *NOAA-12 undergoing outgassing during 1994. During night-time passes in summer, the low level of illumination is illustrated in (b)*

vapour, using the *TIROS Operational Vertical Sounder* (*TOVS*). This supplements the measurements made by balloons carrying radiosondes. The term 'vertical profile' refers to the temperature changes measured at different heights. A *Microwave Sounding Unit* (*MSU*) makes measurements in the oxygen band in the spectrum.

As well as carrying American equipment, there are contributions from Britain and France. The *Stratospheric Sounding Unit* was developed by the British Meteorological Office. It measures radiation emitted by carbon dioxide (CO_2) at the top of the atmosphere. France provides ground station facilities and the *ARGOS data collection system*, which receives weather data directly from buoys, balloons and remote weather stations. The polar orbiters' ARGOS system is particularly useful for those areas not monitored by the geostationary satellites – the north and south poles.

The *Space Environment Monitor (SEM)* records the effects of solar activity, which are then used to predict conditions in the ionosphere. This has a significant effect on radio propagation and radiation levels near the Earth.

Equipment for the *Earth Radiation Budget Experiment (ERBE)* measures different energy levels in the upper atmosphere and provides monthly average figures which help to determine the balance of energy received and radiated by the Earth.

Digital tape recorders are carried by recent NOAA satellites to allow recording of remote areas. Data is later recovered by Command and Data Acquisition (CDA) stations.

The Royal Aircraft Establishment at Lasham is responsible for the acquisition of meteorological satellite data for the Met Office, including high-resolution data.

Emergency beacons

The polar satellites from NOAA-8 onwards, carry a payload which can receive transmissions from emergency beacons (distress signals) and can relay these signals to particular ground stations – *SARSAT* – Search and Rescue. From Doppler measurements made on the beacon signal, the location of the emergency transmitter is calculated and passed to the Mission Control Centre, which then alerts the Rescue Co-ordination Centre.

These transmitting devices are called Emergency Position Indicating Radio Beacons (EPIRB), and when activated, they transmit a beacon signal on one of the following frequencies: 121.5, 243, 2182 MHz. People and groups engaged in journeys to remote areas should consider carrying these devices.

By international agreement some Russian satellites also carry compatible equipment – called the COSPAS system.

Note. If you should ever hear a transmission on these frequencies *do not transmit a response* – you should call the Coast Guard service and provide details.

NOAA beacons

As well as transmitting a.p.t. on either 137.50 or 137.62 MHz, the NOAAs have beacons operating on either 136.77 or 137.77 MHz. This beacon is easily heard, and is used to disseminate data from on-board instrumentation. The data is multiplexed (see later) and includes housekeeping information (described previously in the UoSAT-2 section), and attitude (pointing) data.

Other TIROS Information Processor (TIP) data includes the spacecraft identification, a counter for the minor frame (see later), solar array telemetry, HIRS/2 data, SSU, SEM, MSU and DCS data – previously described.

Multiplexing

While a satellite is in active service, many of its on-board systems are monitored carefully. Some parameters, such as battery voltages and temperatures, change quite slowly with time, so there is little point in measuring their values every tenth of a second! Consequently, although a.p.t. information requires continuous, high-speed measurements of albedo, many slowly varying parameters can be measured perhaps once per second, or even more infrequently. This means that a whole system can be monitored, and the relevant measurements incorporated into the telemetry read-out, just once per telemetry frame. The process of incorporating different parameters sequentially into a relatively long frame rate is called multiplexing.

An a.p.t. frame lasts 64 seconds. A TIP (major) frame lasts 32 seconds, and includes 320 minor frames.

Support from EUMETSAT

For some years the Centre de Météorologie Spatiale (CMS), in Lannion, France has been providing support for the NOAA polar orbiting satellites of the USA. This support consists of controlling the satellites during 'blind' orbits, when the satellite cannot be seen from USA ground stations; data is transferred back to the USA for processing. Overall responsibility for this has now been transferred to EUMETSAT, although CMS continues to provide the actual service.

This is an indication of the interest EUMETSAT has in the continuity of meteorological data from polar orbit, as well as in the continuation of the Meteosat geostationary satellite system. Polar satellites provide essential global data coverage and are used extensively by EUMETSAT member states. Consequently EUMETSAT is exploring ways in which this nominal contribution to the polar system can be made more substantive through a more active contribution to be established during the 1990s.

NOAA 13

This ill-fated satellite suffered from a battery short-circuit, cutting its life dramatically. Consequently few pictures were received from it.

The latest status

NOAAs 12 and 14 are currently the only two NOAA weather satellites transmitting a.p.t. imagery. NOAA-12 continues on 137.50 MHz and NOAA-14 remains on 137.62 MHz.

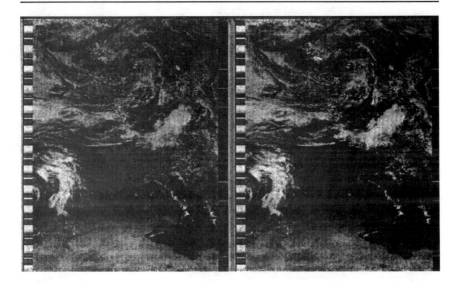

Figure 3.5 *This picture shows the standard image transmission format seen from all NOAA weather satellites during the early days of their lives. Channels 1 and 2 are transmitted during the initial thermal stabilization of the satellite. This process usually lasts for several days while outgassing (see earlier in this chapter) takes place. The effect of the small sensor frequency difference between channels 1 and 2 is seen, when comparing both images*

NOAA 14 – and beyond

This latest in the American constellation of weather satellites was launched on 30 December 1994, into a lunch-time daylight ascending orbit. Within hours of launch, its a.p.t. telemetry was activated and monitored here in Plymouth.

Looking further ahead, two additional advanced meteorological satellites are planned. The satellites are to be designated 'N' and 'N-prime', and are scheduled to be launched after the year 2000. The whole programme is a NOAA-funded cooperative effort between NASA and NOAA, Britain, Canada and France. Goddard Space Flight Centre manages the spacecraft from procurement, through launch and initial in-orbit operations, after which spacecraft operations are managed by NOAA.

EUMETSAT's plans

Plans for a European polar orbiting weather satellite system are under way. The next generation of satellite observation systems, currently at the definition stage, will provide more accurate data with enhanced vertical and horizontal resolution. Such a system is EUMETSAT's Polar System (EPS) which is planned as the European contribution to a joint USA/EUMETSAT satellite programme. EPS is planned to provide observations from the morning orbits of a series of Sun-synchronous, polar orbiting satellite system, while the USA continues coverage from the afternoon orbits.

The programme is expected to start in 1996, leading to the launch of the first satellite in 2000/2001. It (the programme) is based on three METOP (METeorological OPerational) satellites carrying the observational instrument payload, and includes procurement of launchers, and development of the ground processing facilities and operations. Launch dates in 2005 and 2009 are foreseen for subsequent launches, providing coverage until at least 2014.

The orbit is expected to be Sun-synchronous at about 830 km, with a mean local solar time equator crossing for each orbit pass, of 09:00 hours. Instrumentation is intended to be compatible with the latest NOAA satellites, with the addition of an advanced infra-red sounder, microwave sounding and a complementary climate payload.

Convergence of NOAA and DMSP programmes

The National Oceanic and Atmospheric Administration (NOAA) satellites, and those operated by the Defense Meteorological Satellite Program (DMSP) will be merged. The USA's National Performance Review called for the converging of the two programmes in order to generate cost-savings and reduce the duplication of effort. The convergence was approved by President Clinton on 5 May 1994.

The basic system calls for three polar-orbiting satellites spaced at 60° separation, having northbound equator crossings at 0530, 0930 and 1330 local time. The full implementation of this project is not expected before 2004.

4 Weather and resources satellite systems

The Commonwealth of Independent States

The CIS continues its weather satellite programme for the same reasons as America – the need to map remote regions, the search for resources, and routine monitoring of severe weather. Their responsible organization was the USSR Research Centre for Earth Resources Exploration, which is under the State Committee for Hydrometeorology. Their data receiving and processing centres are based at Moscow, Novosibirsk, Khabarovsk and Tashkent, with over 80 simplified receiving stations elsewhere.

From the end of the 1950s, up to the mid-1980s, information on the (then) Russian weather satellite scene was very difficult to obtain because of the intense secrecy in which the former USSR kept its satellite programme. My first enquiries of the Russian embassy, made in 1966, in my capacity as a young space enthusiast, eventually resulted in some rather dated booklets being sent to me, which contained little real information.

From the start of the Gorbachev era, details were increasingly released to western groups. My enquiry of the Russian Meteorology Service, made during the late 1980s, produced a most helpful booklet on their METEOR (weather satellite) and OKEAN (oceanographic satellite) series, from which some of this information originates. Dr. Michael Zakharov has kindly provided the following information for inclusion in this chapter.

During 1995, new facilities were nearing completion; a NOAA HRPT acquisition station was installed at SMIS Laboratory (Space Monitoring Information Support Laboratory, of Space Research Institute, Moscow, Russia) and was connected to the Internet (Russian Space Science Internet) via the IKI local area network. This station covers the whole of Europe, Russia (the European part and western Siberia), Ukraine, Belorussia, Middle Asia (Kazakhstan), Middle East (northern part). The telemetry data acquired is being archived on a Sun workstation for future access and processing. The acquisition station and PC-based workstations at SMIS are equipped with the full range of original data processing software, from data extraction and AVHRR multi-channel processing up to the final geographical referencing and general-purpose image processing. Acquired data is expected to be used mostly for the purposes of weather observation, pollution detection, forest fires detection and other tasks of regional monitoring.

Early Russian resources (oceanographic) satellites

The first Russian experimental craft, used for collecting Earth resources data was launched in July 1974. This was part of a series which became known as METEOR-PRIRODA, and acquired multi-channel information. The first two satellites in the group were put into orbits about 900 km high, having inclinations about 82°. Launch names remained 'COSMOS', the name used for almost all Russian satellites regardless of type.

A number of these COSMOS satellites were seen to transmit pictures using the a.p.t. format – described in Chapter 3. COSMOS 1500, launched on 28 September 1983, carried equipment for observing the weather, and occasionally transmitted a.p.t. on 137.40 MHz. This satellite carried an X-band (30 mm) sideways looking radar having better than 2 km resolution, a UHF spectrometer using several wavelengths with varying resolutions, an opto-mechanical four-channel scanner, and receiving equipment to monitor remote sensing stations for further re-transmissions. There was also a data recording facility for later playback to a ground station. I recorded at least one of these playbacks, receiving a clear picture of the North Pole, from an oceanographic satellite passing over eastern Europe!

Until March 1995, this was the only picture that I had ever seen which included the UK. Virtually all transmissions used to happen while the satellites were east of Britain.

The applications

Information provided by these satellites was used to produce new geological maps of the USSR. Short-term events were also monitored; many forest fires were detected by the satellite sensors, enabling action to be taken to put them out.

An assessment of ice coverage in local lakes and shipping lanes was made by satellite, and used for aiding navigation. Estimates made by the authorities suggested that many thousands of roubles were saved by the timely application of satellite data.

COSMOS 1689, launched on 3 October 1985, could also be identified as a member of this series. Such satellites can be recognized from their orbital characteristics. This new COSMOS was launched into a 600 km orbit (this is fairly low), having an inclination of 98°. Without any advance warning, I received early signals when my scanner 'locked' on 137.40 MHz. The signal

Figure 4.1 *Radar and microwave imagery of part of Britain taken in December 1988*

was clearly a.p.t. but sounded quite different to either METEOR or NOAA satellites.

The OKEAN series

The Russians belatedly revealed that COSMOS 1076 and 1151 had been tested as early ocean monitoring missions. These satellites had orbits typical of ELINT (electronic intelligence) satellites, rather than those normally used for Earth monitoring.

Images of varying quality were received from COSMOS 1500, 1602, 1689, and 1766 during these years. Then in July 1988 a new series of COSMOS satellites was identified when OKEAN 1 was heard for the first time. This series also transmits on 137.40 MHz, scanning at 4 lines per second, as did their predecessors. OKEANs 2, 3 and 4 followed in later years.

During the late 1980s, several COSMOS satellites transmitted pictures containing visible-light images, and sometimes including radar images. A third type of image – one from a microwave sounder – was occasionally included. Because of this, the overall format of an oceanographic satellite picture was unpredictable.

The radar image was often of high quality, with resolution about two kilometres, but rarely lasting for more than a few minutes. Radar systems use much power, so night-time radar imagery tends to be of short duration. The radar looks sideways from the satellite's direction of travel.

One of several picture formats used by the OKEAN series includes a sequence of numbers which can be related to on-board operations. OKEAN 1-07 was launched in autumn 1994, and Figure 4.3 was received on 21 October 1994, and shows part of this number sequence. The incrementing number (1019, 1020, ..., seen upside-down in Figure 4.3) is associated with elapsed time since midnight in Moscow. This image shows cloud over the Black Sea, while Odessa enjoys sunshine, in this visible-light image.

Some examples of images from the OKEAN series are printed here. It can be seen just how variable the picture content of these satellites can be. One can receive visible-light pictures occupying almost the whole frame format, sometimes with the number sequence. More often the frame is split into sections containing the different types of imagery.

During late autumn 1994, I collected numerous pictures, often showing long tracks from the radar and microwave sounder carried by OKEAN 1-07. Figure 4.7 was received on 21 November 1994, during a southbound pass with both radar and sounders operating. The picture starts with soundings of the Gulf of Bothnia, showing the small islands. The long track across eastern Europe reveals considerable detail in the radar image, though little shows from the microwave sounder. The left edge of the picture contains 'piano-key' telemetry.

On 23 November more radar and sounder imagery were received for a short stretch of track near Lake Ladoga, which included a bright area around

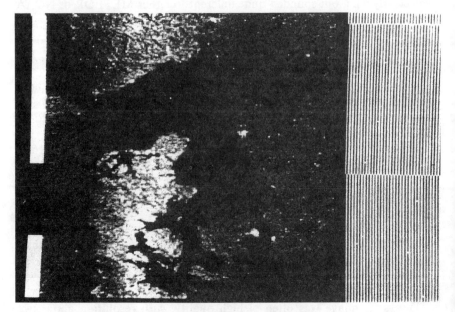

Figure 4.2   *An early OKEAN radar image of Denmark*

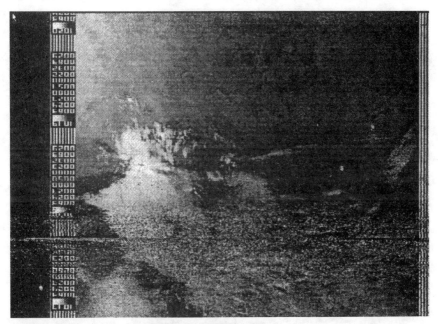

Figure 4.3   *OKEAN 1-7 received 21 October 1994*

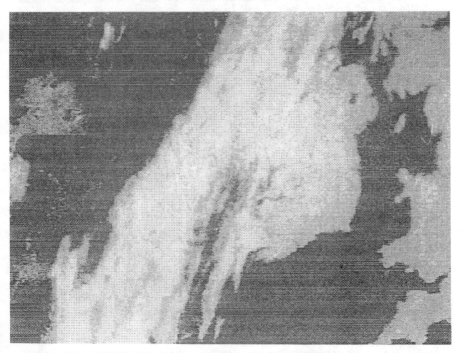

Figure 4.4   *OKEAN 1-7 picture of Britain, 21 August 1995*

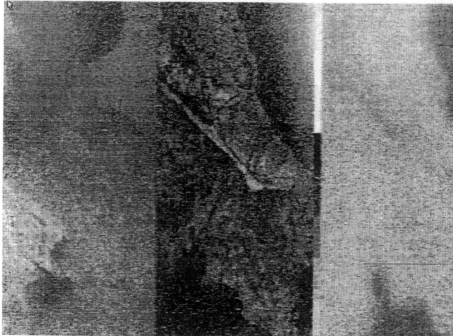

Figure 4.5   *OKEAN 1-7, 23 October 1994*

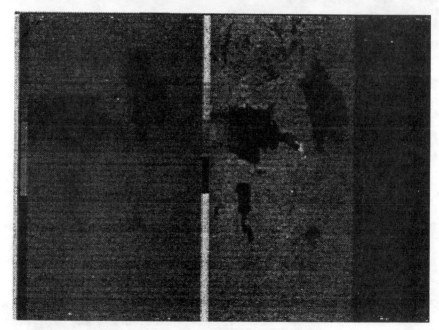

Figure 4.6   *OKEAN 1-07 received 23 November 1994*

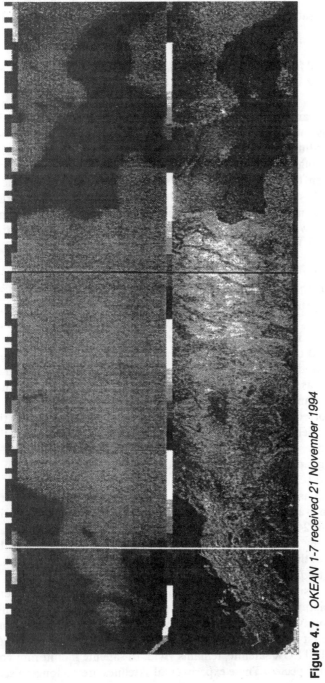

Figure 4.7 *OKEAN 1-7 received 21 November 1994*

Leningrad. The following day (24 November) another long transmission sequence was received, perhaps forming part of a monitoring of this area.

These oceanic satellites remain amongst the most interesting of those that we can monitor.

CIS (Russian) Meteorological Satellites

The METEORS

Although the oceanographic satellites carried some meteorological equipment, as well as multi-zonal scanners with various resolution capabilities, these are power-hungry systems, and have not been used for continuous weather monitoring. That role was taken by the METEOR-class satellites.

Information on the earliest METEOR satellites is scant, but American sources[1] list the first launch as 26 March 1969. It is possible that an earlier COSMOS would have carried experimental equipment prior to this formal recognition of a new class. Further METEOR launches occurred on 6 October 1969 and 28 April 1970. After that date, launches became regular.

METEOR class 1

A number of satellites in this class were launched, including METEOR 1-30, which was in operation for several years. Launched on 18 June 1980, it had an orbital inclination of 97.7° and a period of 94.3 minutes. A study of such an orbit shows that it is Sun-synchronous, like the NOAAs.

METEOR 1-30 transmitted a.p.t. on a number of frequencies, including 137.120, 137.130 and 137.150 MHz. A further series 1 craft was launched on 10 July 1981, and was also considered to be experimental – that is, it did not provide regular transmissions on a fixed frequency.

I was not involved in monitoring the 137 MHz band, while working in the scientific satellite field, so I picked up the signals from METEOR 1-30 for the first time, when my scanner locked on to it on 4 January 1987 at 1135UTC. It was transmitting a.p.t. on 137.02 MHz, at four lines per second (240 l.p.m.), with a resolution better than 2 km. I had not been aware of its existence.

After logging it as an 'unknown' satellite, I sought expert advice. The first suggestion given was METEOR 1-30, but 'not on that frequency' was the comment. Within 24 hours both frequency and satellite were confirmed!

It seems that few people monitored METEOR 1-30, so I passed on the information about the new, lower frequency. Careful monitoring revealed that the transmitted frequency was slowly drifting down.

The picture transmitted contained no phasing bars, or any other markers. There was no picture correction either, so panoramic distortion was quite phenomenal! However, I saw some of the best a.p.t. pictures ever seen, from this craft. That first picture of Italy, passing over at a time when I was not expecting any METEOR satellite, remains vivid. In satellite monitoring you can expect the unexpected. These experimental satellites are no longer used.

Figure 4.8 *METEOR 1-30 image of icebergs around Denmark in March 1988*

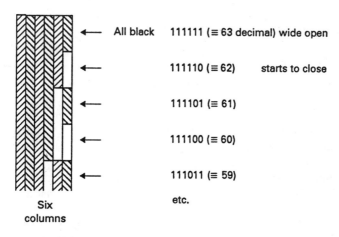

Figure 4.9 *METEOR edge code*

METEOR class 2

During the operation of METEOR-1 series craft, the first in a new series was launched – METEOR 2-1. Later METEOR-2 satellites (from about 2-8) have orbits of average height 950 km, compared with the METEOR-3 series, which came later and are some 1200 km high.

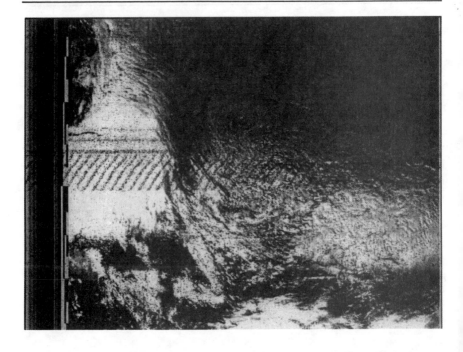

Figure 4.10   *Picture showing aperture changes*

Both satellite classes provided continuous transmissions of pictures during the part of their orbits where they were illuminated by the Sun. Like the NOAA series, not all satellites in a particular class are operating at any one time. The published policy in 1988[2] was to have two or three operating continuously to provide regular pictures for monitoring the distribution of clouds, storms, snow and ice coverage. Temperature profiles, local and vertical, are also monitored.

Launches of these satellites were regular and continued until the launch of METEOR 2-21, in 1993, which appears to have been the last in this series. Launches of the METEOR series have been characterized by the early transmission of slow-scan, infra-red imagery (s.s.i.r.), often heard during the late evening, perhaps a few hours after launch. I was sometimes amongst the first (if not the first on one or two occasions) to hear these birth cries of a new METEOR.

Slow-scan, infra-red imagery used in class 2 satellite transmissions was not of high quality, and rarely lasted for more than a few days. The satellite then reverted to visible-light-only transmissions, switching off as it entered the night.

Picture format of series 2

The visible pictures consist of a.p.t. scans containing picture data, black and white bars (columns) which changed position, a sequence of vertical lines and a grey scale. At one time there were a specific number of the vertical lines, and these could identify the satellite. During recent years, the number has remained fixed. By carefully watching the changing columns, it becomes evident that they are associated with the brightness, or level of solar illumination, of the scene below the craft.

At the start of transmissions, (following a period of eclipse during which there are usually no transmissions,) the black column is at its widest, representing a fully open aperture. Within a few seconds of the craft passing over brighter, increasingly illuminated scenery, the aperture starts to decrease, accompanied by the changing sequence of bars.

These changing bars (technically referred to as edge-code), shown in Fig. 4.9, can be interpreted as a binary indicator. There are six columns, and black is considered to represent a 1, with white as 0. Therefore with the aperture fully open, all six black bars represent the binary number 111111. When illumination increases sufficiently, the first change is a reduction of binary 1 in the far right column, i.e. to 111110. The next reduction is to 111101, and so on until the aperture is fully open represented by 000000. These binary numbers translate to 'normal' numbers (base 10) from 63 to 0.

This indicator has been shown to correlate well with the amount of solar illumination at the point below the satellite (usually called the 'sub-satellite point'). Geoffrey Perry established this analysis.

METEOR class 3

The first in this new series of satellites was launched on 24 October 1985. From a higher orbit (1200 – 1250 km), and inclination of 81° to 83°, a greater swath width is observed. Depending on the resolution used, swath widths between 2600 km and 3100 km are achieved. On-board recording facilities allow transmissions of stored data on command. Quality can be very good, and the published resolution[2] is as high as 0.7 km, though direct (real-time) transmissions normally use lower resolution, even in the highest mode.

Downlink power is about 5 watts, similar to the NOAAs. The 2400 Hz sub-carrier has been more stable since METEOR 3-4, and the modulation level is 90% for peak white. This means that for the visible pictures, an external reference signal is not required. Scan rate (the rotation of the satellite) remains at two lines per second.

Instrumentation carried by METEOR satellites is detailed in Reference 2 listed at the end of this chapter. Table 4.1 summarizes the hardware.

Proton/electron flux in the 0.15 – 90 MeV band is measured by a radiation monitor.

On-board processing corrects geometrical distortion, in which the area

Table 4.1 *METEOR hardware*

Band (μm)	Instrumentation
0.5–0.7	The scanning telephotometer provides direct image transmission. Global survey mode allows larger swath widths.
10.5–12.5	A scanning infra-red radiometer with a below-satellite resolution of 3 by 3 km.
8–12	A scanning infra-red radiometer having a nadir resolution of 10 km in global survey mode.
9.6–15.2, 18	10-channel (multi-spectral) scanning to measure atmospheric temperature and humidity; resolution 50 km for a swath width of 1000 km.

directly below the satellite (its nadir) is seen in greater detail than an area nearer the edge of the field of view.

Future plans include high resolution transmissions in the 1.7 GHz band, as currently performed by NOAA satellites.

Further METEORs in this series were launched during the late eighties and my habit of writing in the evening while the scanner monitors the skies, continues to enable me to log early transmissions.

Winter effects

During winter passes of the METEORs, when they are travelling near the UK, they are coming from, or going towards, a dark North Pole. The satellite is normally off while in darkness – unless transmitting infra-red a.p.t.; at some stage it will therefore cross the terminator (the night–day boundary). This can be heard most dramatically if you tune a receiver to the expected frequency (currently 137.85 MHz), and simultaneously watch the satellite on a good graphical satellite predictions program!

Consider a METEOR class 3 satellite travelling northbound on a westerly pass, during winter mornings. The satellite is on – and therefore transmitting – in the sunshine, but, as it approaches the morning terminator, which will be over the western Atlantic, you can see the aperture indicators open as darkness approaches. The satellite switches off a few seconds after the aperture is fully open – and enters darkness.

The nature of METEOR orbits

To understand the operations of METEOR satellites, we can look at a typical orbit, e.g. that of METEOR 2-21. It has an orbital inclination of about 82°, and a mean motion (the number of orbits completed every 24 hours) of 13.83. This identifies it as a class 2 METEOR. From this figure one can calculate that its orbital period is about 1 hour and 44 minutes. (*Note*: orbital period = 24/13.83, giving us the time in decimal hours (= 1.7353); the decimal part can then be multiplied by 60 to calculate the number of minutes (= 44).)

Using a suitable satellite predictions program, list the daily passes of any selected METEOR class 2 satellite, and watch how they gradually change. During a period of a few days, the passes slip forwards (westerly) in time. So METEOR-2 satellites are characterized by pass times, for any particular location, being some 20 minutes later each day. The plane of their orbits moves westwards, relative to the Sun (i.e. local time). A sample print-out of times illustrates this effect (Table 4.2).

Table 4.2 *Predictions for METEOR 2-21 8 January 1995 (Not every pass is included)*

Date	Event	Time	Event	Time	Maximum elevation
Jan 08:	AOS	2004	LOS	2021	65° east northbound
		2149		2205	26° west northbound
Jan 09:	AOS	0507	LOS	0521	15° east southbound
		2023		2040	77° west northbound
		2209		2223	18° west northbound
Jan 10:	AOS	0525	LOS	0540	21° east southbound
		0710		0727	80° west southbound

Selecting the first pass shown above, we see that on successive days, it should be heard (assuming it was transmitting at the start of each pass) some 18 minutes later each day. It passes further to the west, that is, the previous pass reached a maximum elevation some 65° in the east, while travelling from south to north. Because of the rotation of the Earth, and the small daily progression of the orbital plane, the next day the satellite reaches 77°. Later days see that pass continue to move further westwards, eventually 'vanishing', being 'replaced' by another one starting in the east.

This daily progression of the orbital plane, relative to the Sun, has implications for this class of satellite. The illumination received by the solar panels does not remain the same during each orbit – as occurs with Sun-synchronous satellites. NOAA weather satellites are put into Sun-synchronous orbits – those where the orbital plane of the satellite slowly moves to maintain the same face to the Sun. Sun-synchronous orbits receive a constant level of illumination, which can be beneficial to the life of the on-board batteries.

Because orbits of class 2 satellites are constantly changing their solar aspect (orbital plane with respect to the Sun), watching such a group of passes, you notice they slowly advance towards the following terminator.

Predicting operation changes

Understanding these satellite orbits reveals the limitations of the satellite's operation. With the solar array subjected to constant changes of illumination, there are periods – such as those described above – where the satellite is near the terminator; solar illumination is decreasing rapidly. At such times, the satellite controllers normally switch the satellite off, and re-activate one of the others having a more favourable orbital plane.

A study of later class 2 weather satellites shows that their orbital planes differ widely, so change-over is not a problem. In fact, within these groups, satellites can be identified in sub-groups having similarly orientated orbital planes.

Between 1990 and 1992, operations switched between METEORs 2-19 and 2-20. Predictions of the timing of switch-over were fairly straightforward as each satellite approached the morning terminator.

Other instrumentation

Recent concern for monitoring the ozone layer resulted in class 3 carrying experimental instrumentation to collect data on total ozone concentration and its vertical distribution above the Earth. As with NOAA satellites, many other meteorological measurements are made during a.p.t. transmissions.

During the late 1980s, METEOR-3 series satellites came into increasing activity, and the operational satellite often transmitted continuously. In sunlight, the satellites transmit a.p.t. data showing the usual grey scale with vertical bars. As with class 2 satellites, one set of bars forms the aperture indicator for the scanning camera, and can be converted to a number, when interpreted digitally.

METEOR class 3 Infra-red

As mentioned, class 2 METEORs did occasionally transmit infra-red (i.r.) of limited quality. Infra-red transmitted by class 3 satellites has been of good quality, having the appearance of NOAA i.r. but in reverse – see Fig. 4.12!

Because the transmission contains only i.r., contrast has been adjusted to cover the full dynamic range of the picture's content, so METEOR 3 i.r. pictures are relatively easy to interpret, once you remember that warm seas are white, and cold clouds are black!

FENGYUN: the Chinese connection

China has always been associated with rocketry, so it was only a matter of time before the Chinese announced their proposal to orbit a weather satellite. On 6 September 1988, FENGYUN-1 was launched. I was not aware of the frequency, or even the launch date, but my scanner locked on 137.06 MHz on 29 September 1988, when a sound similar to NOAA-11 was heard. I didn't

Figure 4.11 *Meteor visible-light image of Greenland*

Figure 4.12 *Meteor infra-red image of south Greenland*

receive any identifiable imagery until a few days later on 5 October at 1653UTC, when a picture similar to those from the NOAAs was received. It was later confirmed as coming from FENGYUN-1, transmitting on 137.05 MHz.

Unfortunately FENGYUN-1 did not remain transmitting for very long, apparently due to problems with its stability. FENGYUN-2 was launched on 3 September 1990, and transmitted a.p.t. on 137.80 MHz for a few months before finally failing.

The most interesting features of FENGYUN were the orbit and picture format. The orbit has an inclination of about 99°, and an orbital period of 103 minutes (at its average height of 885 km), producing a mean motion (MM) of almost exactly 14 orbits per day. This brings the satellite over the UK within a minute or so of that of the previous day. This was achieved by very careful launch and injection.

Studying the picture reveals how close the format is to that of NOAA. There are two images side-by-side, though here I show a close-up of just one section. The difference is seen in the edge where NOAA calibration information is replaced by other indicators (see Figs 4.13 and 4.14). FENGYUN also provided high resolution imagery.

FENGYUN channels

Channel   1: $0.58 - 0.68\,\mu$m
2:    $0.725 - 1.2\,\mu$m
3:    $0.48 - 0.53\,\mu$m
4:    $0.53 - 0.58\,\mu$m
5:    $10.5 - 12.5\,\mu$m

Technical details available for the picture format show that the maximum picture modulation is 87%, used for white (clouds).

Sounds of a.p.t.

Listening to the sounds of different a.p.t. transmissions is informative, but practice is required to acquire an interpretive ear. NOAA satellites have an instantly recognizable 'clip-clop' sound. Correlating this with the picture shows that the clip is associated with one half of the picture format (visible), and the 'clop' with the infra-red portion. To be more precise, each 'clip' and 'clop' contains the short synchronizing tone and the longer picture data. The tones precede the relevant picture.

Listening carefully to NOAA signals, when the craft enters the North Polar regions during a winter northbound pass, the change-over from visible picture to infra-red is clearly heard, without any need for a visual demonstration!

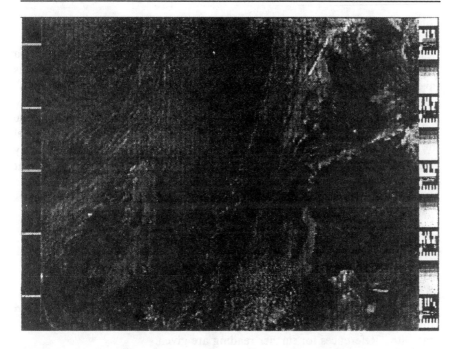

Figure 4.13 *FENGYUN image of Spain – September 1990*

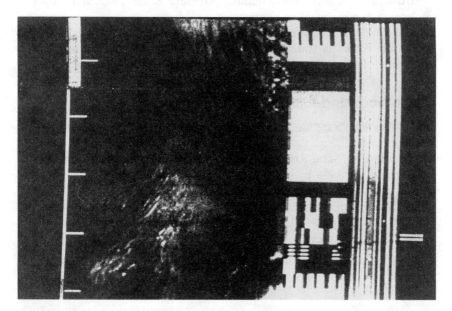

Figure 4.14 *Close-up of FENGYUN image edge*

Similarly, the sound of METEOR-2 visible transmissions clearly reveals the 'croak' due to the phasing bars and grey scale. In between, there is picture data.

During its last years, METEOR 1-30 produced a very individual sound – a burst of 'blowing air' – when it lost its synchronizing bars! All that was left was picture data, consisting of just the modulated carrier – hence the quiet 'puff', which decoded into a clear picture.

Listening to and watching pictures over a period of a few months, one acquires the knack of interpreting the contents without even seeing the picture! When the satellite passes over shower clouds, the change of content is clearly heard to sound like walking on crisp snow – very crunchy. Near twilight regions, the land becomes dark, as does cloud-free sea; this is heard as a low-frequency component of the a.p.t. Just before the satellite enters eclipse, its sound is characteristically dull.

More information

A number of publications have been issued by various organizations concerned with the operation of weather satellites. Some of these provide detailed diagrams of the on-board systems and other measurements made by NOAA satellites. References for further reading are given.

The quarterly magazine produced by the Remote Imaging Group sometimes contains specific information about Russian satellites, written by experts who decoded telemetry from a number of these satellites. These articles (see e.g. Reference 3) are always of interest.

The equipment required to receive signals from those satellites mentioned which are currently transmitting data, is detailed in the following chapters.

References and further reading

1. Publication: 'Satellite Situation Report' issued December 1990 by NASA.
2. Publication: 'State Research Centre for Earth Resources Exploration' published in 1988 by the USSR State Committee for Hydrometeorology.
3. *RIG* magazine number 7, page 19.
4. Publication: *Spaceflight*, Vol. 31, February 1989 – A new eye on the seas, Brian Harvey.

5 Geostationary weather satellites: METEOSAT, GOES, GMS, GOMS, FENGYUN

Since the author and scientist Arthur C. Clarke first wrote about the practical applications of geostationary orbits – where a satellite's period is about 23 hours and 56 minutes (exactly matching the Earth's rotation) – many satellites of different types have been so placed. The distance for this orbit is about 35790 km above Earth, and a ground-based observer of such a satellite sees it appear to 'hover' in the sky because the orbital speed (11069 km/h) exactly matches Earth's rotation. This 'geostationary' orbit is used for all satellites where mission requirements call for continuous contact with the ground station.

Following a decision by the World Meteorological Organisation, a network called the World Weather Watch Global Observation System was set up. International agreements resulted in an allocation of places along the Clarke belt, where various types of satellite can be 'parked' – including a series of weather satellites.

A precise distance from Earth is not the only requirement for a 'geostationary' orbit. The satellite must also have an orbital inclination of zero, in order to keep pace with the rotation of the Earth. If the inclination is non-zero, the satellite will still orbit Earth once per 24 hours, but without remaining over a fixed point – it appears to move in the form of a loop. The American weather satellite GOES-2 has been in this situation since running out of manoeuvring fuel.

International weather monitoring

The constellation (group) of geostationary weather satellites views much of the disc of the Earth, but – because of their positions above the equator – not the polar regions. America operates the Geostationary Operational Environmental Satellite (GOES) system, with at least two satellites usually operating over the USA; Europe, the European Space Agency (ESA), has a number of

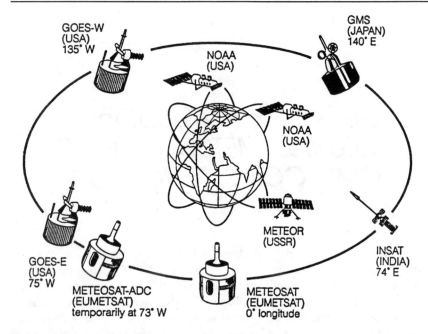

Figure 5.1

METEOSATs positioned near 0° longitude; Japan has the Geostationary Meteorological Satellite (GMS-5) positioned near the Pacific Ocean at 140° east; India has INSAT positioned over the Indian Ocean, and the Common-wealth of Independent States (Russia) has a Geostationary Operational Meteorological Satellite (GOMS) programme which was delayed – launch finally happening in November 1994. The Chinese also plan to operate a series (called FENGYUN) in geostationary orbit. The entire Earth – except the polar regions – will then be monitored continuously.

All the geostationary weather satellites scan the Earth and produce high-resolution imagery (called primary data). A receiving station equipped to receive this particular type of transmission is called a primary data user station (PDUS). Primary data from geostationary satellites complements the NOAA polar orbiting satellites, which also transmit high-resolution images in the same band – about 1700 MHz.

Several other data streams are transmitted by the geostationary satellites, including wefax (a.p.t.-type) pictures, and the latter are transmitted on the same frequency (1691.0 MHz, and 1694.5 MHz in the case of METEOSAT). We shall look at the a.p.t. transmissions in detail.

Because of systems compatibility between most of the geostationary weather satellites, special arrangements were possible when there was a requirement to supplement the American GOES constellation, when the GOES launch schedule fell behind.

METEOSAT – Europe's weather eye

The first in the series of European geostationary weather satellites was METEOSAT-1, launched on 23 November 1977. The satellite was equipped with sensors to monitor visible radiation, thermal infra-red, and water vapour infra-red. The METEOSAT programme was started by the French space and meteorological authorities, but its national status was widened in 1972, when other European countries became involved.

EUMETSAT

EUMETSAT, the European Organisation for the Exploitation of Meteorological Satellites, is an independent inter-governmental organization of 16 European states, which came into being on 19 June 1986. Its primary objective is to establish, maintain and exploit European systems of operational meteorological satellites. Its headquarters is in Darmstadt. From January 1987 EUMETSAT took over responsibility for the continuation of the METEOSAT programme started by ESA, although ESA continued to launch and operate the satellites, and also process and archive the data.

When METEOSAT-1 failed on 24 November 1979, plans were already under way for the next in the series. METEOSAT-2 was launched on 19 June 1981. After the checking out of on-board systems it became fully operational on 12 August 1981.

The next in the series, METEOSAT-3 (called P2), was a re-furbished prototype, but, after launch (15 June 1988), was found to have an antenna problem, causing a reduced-power signal to be transmitted during portions of the satellite's spin. For amateur operations, in which dishes tend to be rather smaller than the officially stipulated minimum size, the effect is that the satellite signal can fluctuate. Problems with this somewhat degraded signal have lessened for the amateur during recent years, with the arrival on the market of low-noise, affordable pre-amplifiers. More is given on this in a later chapter.

METEOSAT-3 was followed by the first in the METEOSAT Operations Programme (MOP) called METEOSAT-4, launched 6 March 1989. This was commissioned, and entered full operations on 19 June 1989, from which time it was referred to as MOP-1. The satellite's two imaging channels were named:

A1 on 1691 MHz
A2 on 1694.5 MHz

Some transmission formats (that is, specific pictures) were transferred from one channel to the other, in order to improve compatibility with GOES transmissions. Some other changes were made, also effective from MOP-1:

Two identical, adjacent, visible-light channels, moved to $0.5 - 0.9$ μm, each channel producing 8-bit data, therefore providing 256 possible levels. The four channels each have a back-up.

Figure 5.2   *METEOSAT*

Orbital changes

Being in geostationary orbit, METEOSAT (and other satellites in geostationary orbits) is about one tenth of the distance to the Moon – and this has its effects. METEOSAT's orbit is above the equator, but the Sun and Moon do not lie in the plane of Earth's equator, so their gravitational effects, combined with irregularities in the earth, tend to pull METEOSAT out of geostationary orbit. The satellite is fitted with on-board gas thrusters to compensate for these changes.

The 'fish' anomaly

In October 1989, an anomaly was detected in METEOSAT-4 image data. The anomaly is caused within the Synchronization Image Channel (SIC) processors which control the radiometer and the sampling, digitizing, multiplexing and formatting of the image. This anomaly occurs only in certain temperature ranges, and is consequently limited to certain periods of the year when these temperatures are experienced. The anomaly reduces images quality by producing randomly distributed horizontal pixel corruptions of between 5 and 35 pixels in length – these are the 'fish'.

During this period of pixel anomalies, for safety reasons, METEOSAT-4 is completely switched off for about 15 days – a period known as the 'fish season'!

The launch of METEOSAT-5 (MOP-2), was unfortunately followed by the discovery of a fault within days of formal operations, causing routine imaging to be transferred back to MOP-1.

Decontamination

In order for the infra-red (thermal) sensors to work properly – that is, to respond to low temperatures on the Earth – they have to be cooled. Using a radiator surface at one end of the spacecraft, a temperature at about 90 K is maintained for these sensors. As a consequence of this colder surface, there is a formation, on the optical surfaces, of condensing contaminants, such as ice, released from the warmer body of the spacecraft.

The satellite is periodically decontaminated by allowing the optics to warm up, therefore vapourizing these deposits. Announcements to this effect are displayed on the METEOSAT administration messages.

Picture generation

Descriptions of image transmission formats from the METEOSAT series are published by various agencies, so I am cautious about duplicating comprehensive notes here. However, a book principally about weather satellites must include information on the formats and availability of METEOSAT pictures!

The Earth is scanned every thirty minutes by METEOSAT, during its 100 revolutions per minute spin, and data in the visible, thermal infra-red and water vapour parts of the spectrum are obtained. After each strip has been scanned (corresponding to one rotation of the satellite), the radiometer's telescope is stepped through a small angle. By this means, a complete, full spectral image is obtained every 25 minutes. Re-positioning of the radiometer – before the start of the next scan – is performed during the last few minutes of the half-hour.

Raw data obtained from this scan is transmitted to, and processed at, Darmstadt Ground Station. From this one data set, two data streams are produced – primary data (PDUS) and Secondary Data (wefax – weather facsimile). For the second data stream, the full disc of the Earth is divided into a number of segments – small sections of the complete image. Land outlines and major intersections of latitude and longitude are added to the image, and can be of considerable help when analysing cloudy images, where land recognition may be difficult. Wefax formats contain reduced resolution imagery, compared with the primary data from which they originate.

The imaging payload is a multi-spectral radiometer which produces images of the complete Earth, as described previously. The infra-red image comprises 2500 picture elements (pixels) per scanning line, of which there are also 2500 pixels. The resolution of the infra-red image (i.e. the minimum size of resolved detail) at the point immediately below the satellite – the nadir, or sub-satellite point – is 5 km. Each visible-light image comprises 2500 lines of up to 5000 pixels, and the channels can be operated together, giving a better resolution of 2.5 km, in the east–west direction. The visible (VIS) channel can operate with one sensor, so such formats contain reduced resolution – 5 km. To maintain picture symmetry, each second line in these formats is duplicated. Primary digital data is routinely transmitted on 1694.5 MHz at pre-set times, every 30 minutes, with other formats transmitted at other scheduled times.

Primary Data

The highest-resolution picture data available from METEOSAT is that from the primary data stream, transmitted on 1694.5 MHz. Reception requires a minimum dish size of 1.6 m and high-quality amplifiers and receivers. Only within recent years has such hardware become available to the amateur community. I took delivery of the first commercial unit produced by Timestep Weather Systems of Wickhambrook and have obtained primary data images since – apart from mechanical problems caused by dish mountings ageing rapidly in the West Country weather.

PDUS images from METEOSAT are transmitted in various formats – full-disc and part-disc images. Once each day, at 1134UTC, the whole disc, maximum resolution AV format is disseminated – taking almost 25 minutes. The A prefix means a full-disc image; so

- AV indicates visible-light image of the whole disc
- AW indicates full-disc water vapour image
- AIW indicates full-disc infra-red and water vapour
- AIVH indicates full-disc infra-red, interlaced with half-resolution visible
- B represents the European sector
- BIV indicates infra-red and full resolution visible image
- BIW indicates infra-red and water vapour images
- BW indicates water vapour
- BIVW indicates the three interlaced images.

This image in Fig. 5.3 was relayed to METEOSAT-5 for re-transmission within the schedule, under the format name LXV. The number of such re-transmissions is limited by numerous time constraints, so resolution is reduced to allow more image transmissions.

WEFAX image format

Wefax pictures, transmitted by METEOSAT and GOES, consist of four-minute frames, each originating from the high-resolution scan taken every 30 minutes. Several frames are repeated at regular intervals. We now look at the format of a typical image, shown in Fig. 5.6.

Figure 5.3   *This shows a typical whole-disc, visible-light image, originally scanned and produced by METEOSAT-3, from its position near the eastern coast of the USA*

Images from METEOSAT are suitable for hardware (framestore) or computer processing. Each line of the picture lasts for 250 ms (ms denotes millisecond – one thousandth of a second), so there are four picture lines per second. As seen in the diagram, the image starts with a 300 Hz tone lasting 3 seconds. Hardware or software can detect this tone, and the picture can then be analysed, line by line. The second portion of the signal contains 5 seconds of phasing tone, itself containing 12.4 ms of black bar. This forms the edge of the frame.

The bulk of the frame (200 seconds plus digital header) contains a white edge (the line start signal, lasting 11.9 ms) and the picture data, making a total of 800 lines of actual picture. The METEOSAT header forms part of the image and is the small inset identifying the format (C02 etc.). The last part of the frame contains 5 seconds of 'stop' signal – a 450 Hz tone. This informs the decoder (or

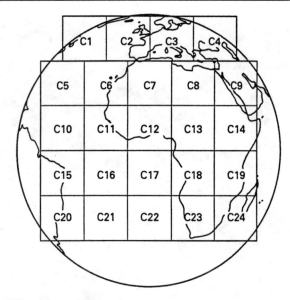

Figure 5.4 *METEOSAT C formats*

software) that the picture has finished. Each image is identical in format (but not content) to every other wefax image.

After each METEOSAT frame, the following 27 second period may include an occasional burst of irregular 'noise', the data collection platform (DCP) messages.

Visible-light images are prefixed with a C, numbered from 1 to 24 – see Figure 5.4 – though not all images are actually transmitted. Infra-red (D) formats are numbered 1 to 9, and water vapour images are E formats, in which the areas are identical in coverage to D formats – see Fig. 5.5.

The main visible images from METEOSAT, are C02, which includes most of Britain, and Spain. The C03 format covers much of Europe, Italy and Norway. A second type of visible frame is also transmitted, using the 'D' format, e.g. C2D.

Monitoring METEOSAT WEFAX images provides enough information to generate a fascination with this ability to see most of the near side of our globe. Using C02 you can monitor our weather and anticipate changes. Pictures are normally transmitted within 30 minutes of the scan, so no-one can receive them before you, not even your local weather centre! They may even receive them indirectly, via a facsimile service – and probably later.

Some examples of METEOSAT WEFAX pictures are shown in Figures 5.7–5.11.

Weather conditions can be monitored 24 hours per day, 365/366 days per year using thermal and water vapour infra-red scanners; visible light images are limited to those times each day when solar illuminations levels are high enough.

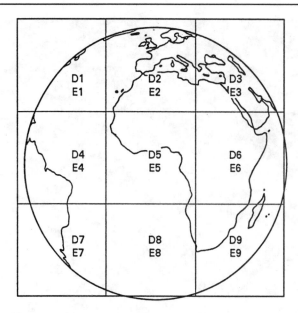

Figure 5.5 *METEOSAT D and E formats*

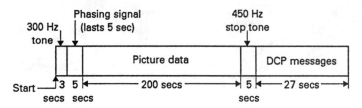

Figure 5.6 *METEOSAT WEFAX image format*

Media interest

With equipment capable of receiving valuable data, there could be interest shown in your work. Data from weather satellites must not be sold or given to anyone for commercial use. By that, I really mean the media. Satellites are funded by meteorological authorities, and amateurs are permitted to receive this data on the strict understanding that it is for their own use or interest. You must not sell weather pictures.

METEOSAT primary data is being encrypted, during a phasing-in process that started in 1994. WEFAX data is scheduled to remain un-encrypted, but will undergo fundamental changes in the next decade–see later.

Figure 5.7   *This is a whole-disk, visible light format (CTOT) image taken on 14 December 1994 at 1200UTC. It illustrates the northern winter; countries to the north of the equator receive less solar illumination than those in the southern hemisphere*

Schools' Projects

A chapter has been devoted to various projects made possible using imagery from the weather satellites. METEOSAT imagery forms a considerable part of these projects.

The pictures

You have to study images closely to appreciate their content. Using a properly designed system, as described in later chapters, you should be receiving

METS 05 OCT 1994 1200 VIS1+2 CTOT

Figure 5.8 *This was recorded on 5 October 1994, a few days after the autumnal equinox, and shows declining levels of solar illumination in the northern hemisphere. Country outlines are added by software at the ground station, and form part of the WEFAX image*

pictures containing up to 64 grey levels – the exceptions being water vapour images, and administration messages.

Sea should appear black, apart from solar reflections. Clouds may show a gradation from white to dark grey – the whitest are generally the coldest, because they are the highest. Darker clouds tend to be warmer, often rain-bearing.

By studying cloud types you may learn to identify the nature of the weather as it approaches Britain. You may notice another feature – clouds seem to hug the coastal regions. Monitoring the C03 frame, which includes Italy, often shows long banks of cloud, tending to lie along the coasts. This tendency is seen in many other areas.

Europe's northern inland lakes form another source of great interest. Some

Figure 5.9 *This is a similar image, but taken on 4 October 1994 at 1800UTC, using the infra-red scanner. Subtle thermal changes can be seen; South America is warming up during its morning; the deserts of North Africa are cooling down as evening progresses*

freeze over during late autumn, and the ice extent can be monitored during winter.

Re-transmissions

METEOSAT not only transmits pictures obtained from its own scanners, but, at regular intervals, also transmits frames originating from GOES and GMS.

METEOSAT-3 (see later) was, until recently, positioned at 70° longitude from where it assisted the American weather-watch programme. Those slots labelled LZ, LY and LR were collected from METEOSAT-3 and re-transmitted from METEOSAT-5 several times per day. Even if your METEOSAT dish (or yagi)

Figure 5.10 *ETOT-1 is the infra-red, water vapour band image, recorded at the same time as the previous image, but indicating concentrations of water vapour, rather than heat*

cannot be easily pointed westwards towards METEOSAT-3, you can monitor the larger part of America by collecting these images.

Tropical areas covered by METEOSAT-3 images are the breeding grounds for hurricanes, and are worth regular monitoring. Each frame taken by METEOSAT covers an identical area to the last equivalent frame, so it is possible to record a sequence of such frames. These can be viewed in quick succession – an animation sequence.

METEOSAT data

Many types of data are produced from the METEOSAT scans. The official product list includes:

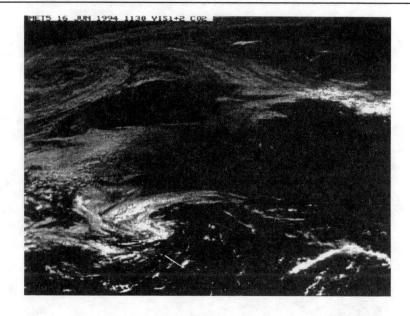

Figure 5.11   *A typical summer C02 format image showing a well-illuminated northern hemisphere. A high pressure area can be assumed to be stationed near France, extending over the eastern Atlantic Ocean*

- Cloud motion vectors (winds)
- Sea surface temperatures
- Cloud analysis
- Climate data
- Humidity in the upper troposphere
- Cloud top height measurements

METEOSAT – the future

EUMETSAT plans to continue the METEOSAT satellite system. The pre-operational phase used numbers 1 and 2, the latter finally removed from orbit in December 1991, after 10 years of service. METEOSAT-3 has been operating from a position over the eastern coast of America. METEOSAT-4 (MOP-1) operated until 1994; METEOSAT-5 (MOP-2) is the currently operational satellite, with number 6 in position near 5. These satellites provided a continuous operational meteorological system until the end of the METEOSAT Operational Programme (MOP) in 1995.

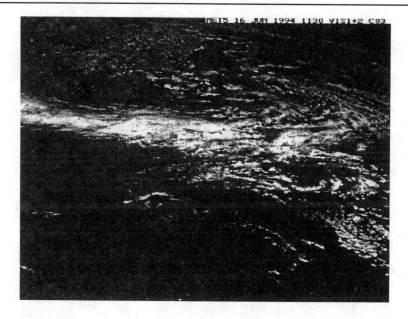

Figure 5.12 *The adjacent area to the C02 format shows Europe on 16 June 1994. Large areas of land are bathed in summer sunshine, and the Mediterranean Sea is clear of clouds*

Looking beyond this first series, EUMETSAT is working with ESA to create a second generation of METEOSAT (MSG) which is planned to come into service within a few years. A METEOSAT transition programme (MTP) has been initiated to facilitate the launch of a further METEOSAT should one prove necessary. METEOSAT-7 is under construction by Aerospatiale in Cannes, France and will be launched by EUMETSAT at a suitable time after mid-1996. MTP is scheduled to start at the end of 2000.

Meteosat second generation

Studies regarding the Meteosat second generation (MSG) continue by both ESA (for the space segment) and EUMETSAT (the ground segment). It has been agreed that MSG should have a new spin-stabilized design, rather than the previous three-axis stabilized satellite. This has important consequences for the mission. The primary objectives call for an 8 or 10 channel radiometer producing images every 15 minutes. Secondary objectives call for monitoring of the climate and environment.

The detailed design and development phase is expected to start in July 1995,

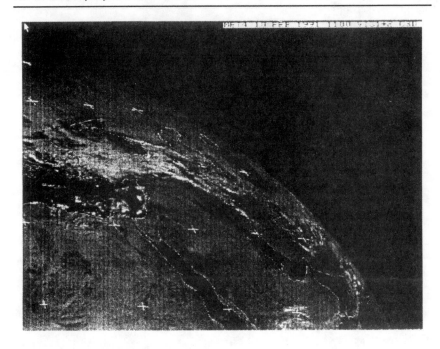

Figure 5.13   *METEOSAT C3D on 10 Feb 1991 showing smoke blown south-east from Kuwait*

with 2000 as the planned launch date for the first MSG satellite. The EUMETSAT programme also calls for the manufacture of a further two satellites, the procurement of launch vehicles for all three satellites, and the development of the complete ground segment and operations of the system until 2012.

MSG Instrument Details

The current draft specifications are indicated: the MSG will be drum-shaped, approximately 3.2 m in diameter, weighing 1750 kg at launch. Each satellite will be positioned in geostationary orbit over the meridian for a design lifetime of 7 years.

The MSG radiometer is considerably different from the MOP radiometer. It will scan the Earth in 12 different spectral channels, and is called a spinning enhanced visible and infra-red imager (SEVIRI). In 11 of these channels, images will be composed of 3750 lines of 3750 samples, corresponding to an on-ground resolution of 3 km. The 12th channel will be dedicated to

Figure 5.14 *Transmissions of images originally obtained by the Japanese satellite GMS-4, started routinely (from METEOSAT) on 12 August 1993, with the sequence GMSA, GMSB, GMSC and GMSD infra-red images. Each is one quadrant of the disc seen from GMS, and is transmitted on channel A2, which contains all re-transmitted images. These images provide a constant source of images of tropical storm areas, and are unsurpassed for school project investigations–see Chapter 9*

high-resolution visible (HRV) imaging at 1 km resolution (marginally better than NOAA h.r.p.t.!). As mentioned, images will be received on-ground every 15 minutes.

The MSG satellites include three transponder channels, allowing downlink of the raw images and satellite telemetry to the primary ground station, relay of other data from ground or air-based environment monitoring stations (data collection platforms – DCP) and the re-transmission of processed imagery and other data to users in high rate and low rate data stations.

Hardware and software for MSG

It is evident from the vastly different specifications for METEOSATs MOP and MSG that equipment now used for primary data and WEFAX imagery is unlikely to be of use with MSG telemetry. At the time of compiling this chapter, information on the possible dual transmission of WEFAX and PD

Table 5.1 *Proposed MSG basic imagery mission*

Group	Name	Channel (µm)	Sampling interval (km)	Main purpose(s)
	VIS 1	0.60–0.67	3	Cloud detection
VIS & NIR	VIS 2	0.77–0.83	3	Convection monitoring
	IR 1.6	1.53–1.70	3	Precipitation monitoring Cloud motion winds (CMW) Surface radiation fluxes
	IR 1	10.3–11.3	3	Cloud detection
WINDOW IR	IR 2	11.5–12.5	3	Cloud height assignment
	IR 3.7	3.5–3.9	3	Wind extraction Surface temperature
	WV 1	5.8–6.7	3	CMW height assignment
WV IR	WV 2	6.9–7.3	3	WV winds Stability analyses
HRVIS	HRVIS	TBD	1	Convection monitoring Fog monitoring

was not available. I would anticipate a sharp transition from the current system to the new format during the next century.

This chapter looked at the satellite and its provision of imagery. A later chapter looks at the equipment necessary to receive and decode METEOSAT telemetry.

GOES: the American geostationary weather satellite constellation

Within the international network of weather satellites in geostationary orbit, the American group (constellation) is called the Geostationary Operational Environmental Satellites (GOES). There are several such satellites in position, but the ravages of time and the Challenger accident temporarily halted replacement of the ageing craft.

The nominal GOES mission calls for two satellites to operate above the eastern and western coasts of the USA. One, positioned at a nominal 75° west longitude, is called the GOES 'East' position, and provides coverage from half-way across the United States to the middle of the Atlantic Ocean. The other, at longitude 135° west, provides coverage from the western side of the USA, well into the Pacific Ocean.

The failure of GOES-6 forced NOAA to reposition its sole operating geosynchronous satellite, GOES-7, to a point midway over the continental United States. In August 1991, EUMETSAT loaned METEOSAT-3 to NOAA to supplement Atlantic coverage. Both these weather satellites far exceeded their design lifetimes. GOES-7 is out of manoeuvring fuel, causing its orbital inclination to change each day, while METEOSAT-3 has been removed from geostationary orbit.

GOES-2: This satellite is in a stand-by position near 105° west. Its orbital inclination is about 9°, causing it to loop around our western horizon. For a time, it acted as the 'East WEFAX' broadcaster, providing a.p.t. and digital telemetry. The schedule of broadcasts is shown in the transmission at 1055UTC.

GOES-3: This satellite is also in a stand-by position, at about 175° west, having an orbital inclination of some 8°. Its location means that it cannot be monitored from the UK.

GOES-6: This was in the 'Goes-West' position at about 135°, with an inclination of some 3°. It was broadcasting WEFAX and other information.

GOES-7: This was the prime satellite, located at longitude 104° west. It has been moved to 135° west, and obtained imagery for dissemination (transmission) by the other satellites. It was retired in January 1996.

GOES-8: The first test photographs from this GOES weather satellite were transmitted back to Earth on 9 May 1994. The satellite, designated GOES-8 after achieving orbit, is the first in a $1.1 billion series of five next-generation weather satellites known as the GOES-NEXT series. Following a pre-dawn launch from Cape Canaveral Air Station on 13 April 1994, GOES-8 underwent a six-month period of extensive testing.

This new satellite was more than five years late in reaching orbit, so the visual images produced by GOES-8 came not a moment too soon for anxious officials at the National Oceanic and Atmospheric Administration (NOAA) and the National Weather Service.

A GOES-8 launch failure would have been catastrophic, but weather officials had developed a set of 'No GOES' contingency plans just in case. The emergency plans included expanding use of NOAA's polar-orbiting TIROS satellites and commandeering images from a variety of Department of Defense weather spacecraft.

Costing $220 million apiece, GOES-NEXT are the most expensive and sophisticated meteorological satellites ever developed. Space Systems/Loral is the prime contractor for the 4640 pound (weight) spacecraft, designed to supply sharper and brighter images of weather systems to meteorologists. Enhanced clarity is accomplished by increasing the sensitivity of the imaging instruments to recognize 1024 different shades of gray, in contrast to the 64 shades now distinguishable.

The technique of fixed attitude stabilization employed by the GOES-NEXT series will be a major factor for the improved image clarity, enabling the spacecraft to relentlessly examine the swirling weather patterns on the Earth below. Previous weather satellites were spin-stabilized, rotating on their axis like a top, photographing only a thin strip of terrain on each revolution.

GOES imaging

GOES-NEXT satellites are equipped with two instruments; an imager, which will provide visual and infrared images, and a sounder, which will survey the temperature and moisture profiles of weather systems. The imager is a multi-channel instrument, designed to sense radiant and solar-reflected energy from sample areas of the atmosphere. Spectral channels simultaneously sweep east-to-west along a north-to-south pass, employing a two-axis mirror scan system. The imager generates images of an entire hemisphere and sector images containing the edges of the Earth. A revolutionary flexible scan or 'zoom' feature permits imaging of smaller areas, allowing for immediate and continuous monitoring of regional weather phenomena (mesoscale) and detailed wind determination.

Every fifteen minutes, the imager produces visual and infrared images that measure clouds, water vapour, surface temperature, winds, albedo and i.r. flux, as well as fires and smoke. Five different spectral bands (0.55–0.75, 3.8–4.0, 6.5–7.0, 10.2–11.2, 11.5–12.5 μm) will be covered.

The 'zoom' feature will enable forecasters to zero-in on areas as small as about 100 km (65 miles) square, ideal for observations of localized events such as hurricanes or tornado-spawning thunderstorms. This capability, coupled with the ability to precisely measure wind direction and surface temperatures, is expected to provide a boon for hurricane trackers and may dramatically improve predictions of hurricane movements.

The sounder can peek into vertical corridors of the atmosphere, sampling air temperature and moisture content at different levels. It is a 19 channel, discrete-filter radiometer, designed to provide measurements of atmospheric temperature and moisture profiles, surface and cloud-top temperatures, and ozone distribution. A multi-element detector array can sample four separate atmospheric columns simultaneously.

Each member of the GOES-NEXT series of spacecraft is also equipped with a trio of instruments known as the Space Environment Monitor system. A magnetometer measures the intensity and direction of the Earth's ambient magnetic field. An energetic particle sensor and associated high-energy proton and alpha detector monitor solar protons and alpha particles that could prove hazardous to both manned and unmanned space operations. The X-ray sensor provides real-time measurements of solar X-ray emissions to provide early warning of solar activity.

Completing the GOES-NEXT instrumentation suite is an image navigation and registration system that will provide and maintain the geographical location of imager and sounder data with pinpoint accuracy. The satellites will also receive and relay data from geographically dispersed ground-based data collection platforms (DCPs). Weather facsimile (WEFAX) data will be continuously relayed to users.

A dedicated transponder on the satellites will pick up and relay distress signals from aircraft or marine vessels as part of the Search And Rescue Satellite Aided Tracking (SARSAT) system.

The GOES-8 spacecraft was placed into an initial orbit about 800 km (500 miles) lower than expected, reducing the life expectancy of GOES-8 by about 6 months, down to about 8.6 years. The National Aeronautics and Space Administration (NASA) developed the GOES-NEXT series and is responsible for placing the satellites in orbit, while NOAA takes over operational control once the satellite reaches its final orbit.

Each of the GOES-NEXT spacecraft is expected to remain operational for just over nine years. GOES-8 was joined in orbit in Spring 1995 by the GOES-J spacecraft renamed GOES-9 when it achieved orbit. The remaining three spacecraft will be launched, on an 'as-needed' basis, into the first decade of the 21st century.

GOES-4, and some later satellites in this series, were built by the Hughes Aircraft Company. The Goddard Space Flight Centre provides project management for the GOES programme for NASA. NOAA reimburses NASA for the cost of the launch vehicle and launch services.

Transmission schedule

GOES has a significantly different transmission schedule than the familiar METEOSAT sequences. During a 24 hour period it transmits selected METEOSAT images; various quadrants of the Earth as seen from the actively imaging GOES, in the usual water vapour, infra-red and visible light; selected regions from the NOAA polar satellites – as produced on a Mercator projection in infra-red and visible; ice and WEFAX charts, mean sea-level charts and many others. The schedule is transmitted at 1055UTC each day, as is TBUS data at 1410UTC for the different NOAA satellites.

METEOSAT-3 (P2)

Due to the late launch schedule for GOES-8 craft (since launched, as detailed above), the METEOSAT-3 satellite was manoeuvred to 70° longitude, over the eastern coast of America. From here it relayed Atlantic pictures which could be monitored from favourable locations in Britain. Following the deployment of GOES-8 during 1994, METEOSAT-3 was manoeuvred east before leaving routine operations.

GMS-the Japanese system

As with GOES images, METEOSAT-5 routinely transmits images obtained by the Japanese weather satellite, currently GMS-4. The nature of the signal format differs from normal WEFAX in using an FM deviation of 150 kHz – considerably wider than that used by the other geostationary weather satellites. Receivers for direct reception of GMS telemetry therefore need a wide-band FM mode.

GOMS-the CIS system

Positions have been allocated for the Russian Geostationary Operational Meteorological Satellites which were due to start operations in 1992. GOMS was finally launched at the end of October 1994. The satellites have two WEFAX channels and two digital channels.

FENGYUN-the Chinese system

The Chinese have not been lucky with their polar orbiters of the FENGYUN series – both have unfortunately suffered stabilization problems. The geostationary satellites are expected to be in operation soon.

Late-breaking news

See the final chapter for news that was released during the later stages of producing this book.

6 The start of the ground station

We now look at the hardware required to receive satellite signals – antennae; cabling, pre-amplifiers and receivers. I have already mentioned the importance of deciding what you want to do before embarking on the project. You may wish to 'test the temperature' first, as suggested previously, by setting up a minimal station, just for monitoring UoSAT. That was the way I chose.

The principles involved in receiving satellite signals are straightforward. Satellites orbit the planet and, depending on the orbit, may rise and set above the local horizon. Geostationary satellites, those above our horizon, can be monitored. The closer orbiting satellites require antennae suitable for their transmitted frequencies.

UoSAT-2 – a good starting place

UoSAT-2 is a close-orbiting satellite, data from which can be easily received. Equipment similar to that for UoSAT-2 can be used to monitor a range of satellites operating in the band between 136 and 146 MHz. UoSAT-2 transmits telemetry on 145.825 MHz (apart from other frequencies which I shall ignore, simply because this one provides the data in which we are interested). To tune to this band, start with a simple dipole. Later, I shall discuss the benefits of crossed dipoles, but for UoSAT-2, a simple dipole may be more appropriate – see Fig. 6.1.

The length of a simple dipole is related to the frequency at which you wish to operate it. Using the relevant equations (detailed in the next chapter), the optimum overall length of a suitable dipole (one tuned to 145.825 MHz), can be calculated, resulting in the figure 96.52 cm. For use in the weather satellite band, we would want to cut the dipole for 137 MHz. Such an antenna has an average impedance of some 70 ohms; to connect to a receiver, cable of this impedance can be bought – it is the type used for television signals.

The study of antenna theory is a fascinating one, and you can design an

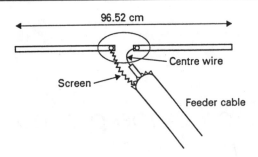

Figure 6.1 *Dipole for UoSAT-2*

antenna for a specific purpose. Adding more elements, above and below, to a simple dipole drastically changes its characteristics, and therefore its impedance.

For test purposes, the dipole can be home-made. Use whatever metal wire is available. I used copper wire but found that thicker strands were needed. Radio rallies are an excellent source of copper wire.

Jargon

Starting off in radio and electronics can provide a baffling amount of new terminology. Matters of impedance matching and different types of connector can leave one feeling overwhelmed. I am therefore striving to avoid unnecessary complications. There will be a time when improvements to the system are sought, and that will be the opportunity to learn more about jargon.

Cables and antennae

Cable construction

If you are unfamiliar with cable connections, the following may help. Cable consists of an inner core to conduct the signal, a central insulating layer, and an outer sheath to protect (earth) the inner signal from external interference – see Fig. 6.2. The satellite signal (145.825 MHz) is received by the dipole as soon as it is within reasonable line-of-sight of the antenna. (A better technical description of the process would be to say that the signal transmitted by the satellite causes the antenna to resonate electrically at this frequency, and, because its length has been carefully cut, it will respond well, achieving a good signal strength.)

The signal will be attenuated (reduced in power) by the electrical energy used to travel along this inner core, so the shorter this cable can be kept, the stronger will be the final signal.

The middle section – that material between the central copper conductor and the outer copper sheath – is called the 'dielectric insulator'. Its composition varies with the cost of the cable. High-cost cable should have good quality dielectric, possessing good characteristics for the specified transmission frequency. The main characteristic is impedance, and the design and structure of the insulator will affect this.

The signal carried along the central core is protected from extraneous electrical noise by surrounding it with conducting material (a copper sheath) connected to a good earth – hence the reference to this outer sheath as the 'earth'.

Were the outer cable not so connected, the cable could pick up interference (noise) from nearby electrical equipment, and the combined 'signal' (including noise) would arrive at the receiver. In practice, other unwanted signals may still be received, but, where good quality cables are used, probably not an excessive amount.

Every dipole receives signals (resonates) over a range of frequencies near the optimum, and will receive further ranges of frequencies that are harmonics (multiples or sub-multiples) of this frequency.

Using my own UoSAT receiver, I experimented with a number of cables and antennae, and found reasonable signals could be received using feeder cable lengths up to 20 metres or so. Try to keep the feeder cable length to less than this distance.

Crossed dipoles

Although it is feasible to receive good signals from UoSAT-2 using a simple dipole, the crossed dipole is preferred, and is almost essential for some satellites.

When a satellite is placed in orbit it must be stabilized, otherwise it will continue with its last imparted motion (the effect of such motion is called a tumble). The satellite is designed to transmit telemetry in the direction of Earth, so some form of stabilization is provided. UoSAT-2 spins on its main axis, and, by suitable spacecraft antenna design, provides a (left) circularly-polarized signal. The signal strength received by an Earth-based dipole therefore changes, as the satellite's antenna changes its relative position. The resulting signal from the dipole can change by 3 dB (i.e. may halve or double in power).

Some antennae sold specifically for UoSAT-2 are described as 'left-circularly polarized'. I bought one some years ago and it appeared to be rather mediocre when compared with a simple dipole; I found the latter gave a better signal strength, and for longer periods. Do not assume that better results will automatically be obtained from a crossed-dipole antenna – experimentation is needed. However, for the weather satellites you will undoubtedly find that a crossed dipole does give a stronger signal.

Antenna phasing

When constructing a crossed dipole, it is essential to connect the two dipoles together correctly – a process called antenna phasing. The radiation (signal) received from the satellite is polarized, and the changing signals induced in each dipole must be electrically added – the phasing harness does this by having a precisely cut length of a specific type of cable. Instructions for connection are provided by the manufacturer – and at least one published the wrong information!

Other suitable antennae

For the UoSATS, and the weather satellites, a simple dipole will receive a signal good enough for initial tests. Weather satellites provide a considerably stronger signal than UoSAT-2, and the (right-circular) crossed dipole is definitely beneficial for these. For general monitoring purposes, a variety of antennae can be used, particularly if there is no intention to process the data.

It is quite feasible to use dipole clusters, as shown in Fig. 6.3, to receive signals in this band. I use a discone mounted in the loft, and receive signals across a wide radio spectrum. Appreciate the phrase 'horses for courses'! With an antenna tuned to a specific frequency (UoSAT-2) you will get the optimum signal. With a wide-band antenna such as a discone you will receive signals over a large spectrum and these are unlikely to be of a suitable quality for data processing.

The addition of more elements (reflectors) to your dipole, increases the gain (the ratio of received signal to that of the simple dipole). I added one set of reflectors to my UoSAT dipole and noted an increase. I also noticed that I got a good signal using a weather satellite antenna, which of course is cut for the lower frequency band (137 MHz).

Discones are ideal for scanners and general wide-band frequency monitoring. A typical discone can detect signals from 40 MHz up to about 1200 MHz. At these higher frequencies, the efficiency may be quite low, so many discone users connect a wide-band pre-amplifier to the discone. This combination can be fairly effective, but again, the signals produced are really only suitable for monitoring purposes. Their quality and purity may change rapidly, making data processing impractical.

Yagis can be effective for both monitoring and signal reception. Their efficiency depends on construction and the number of reflectors. At UoSAT frequencies, a yagi antenna can be large, yet produce little extra, usable signal. Yagis are perhaps best when used for higher frequencies such as the 400 MHz band, upwards. With more reflectors, the gain and beamwidth of the yagi are increased and reduced respectively, so some form of tracking system may be required. At that stage we are leaving the realm of low-cost systems.

Entering the advanced field of tracking equipment for use at higher frequencies, is an interesting step, so a reference to further reading is provided.

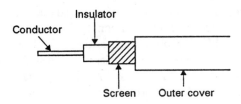

Figure 6.2 *Coaxial cable*

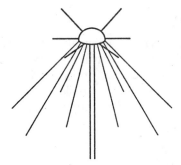

Figure 6.3 *Discone*

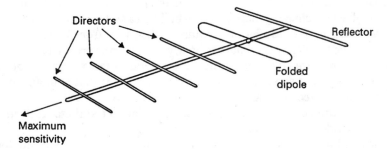

Figure 6.4 *Yagi*

Pre-amplifiers

There is always a temptation to fit a pre-amplifier – a small circuit which amplifies the initial signal received by the antenna. My recommendation, based on considerable experience, is that you should first thoroughly test the antenna at ground level, using the maximum length of cable, estimated to be required. You may find a fair signal can be received just using the antenna alone – feeding the cable directly into the receiver.

If you wish to experiment, connect the pre-amplifier at ground level and do comparison tests. Unless there is a dramatic improvement, don't fit the pre-amplifier. The disadvantage of having a roof-mounted system fitted with a pre-amplifier, is that the antenna is far more likely to receive and amplify unwanted interference. Ground tests may not reveal the extra unwanted noise, later received from a roof-mounted antenna!

Pre-amplifiers of differing qualities are available. Look for one with a narrow bandwidth – so that only frequencies near UoSAT-2 will be amplified. The gain need be no more than 10 dB. Avoid high-gain (20 db or more) pre-amplifiers. You should only provide sufficient gain to overcome feeder losses, and these should be within about 10 db for moderate cable lengths.

Building a pre-amplifier

This construction project is not too difficult, even for beginners, so if you wish to try your hand at building some useful circuitry, there is no better item with which to start. Suitable kits are usually priced within £20 but you may have to provide waterproofing yourself! If you build your own you should try to test the unit separately. Perhaps a friend can provide suitable test gear?

Receivers for UoSAT-2

The UoSAT-2 satellite is an excellent starting place, because of the minimal cost involved in achieving the main aim – building a system capable of actually receiving satellites. Because of the nature of the signal from UoSAT-2, i.e. a narrow-band frequency-modulated signal, we can use any proprietary receiver that can tune to this frequency. There are several suitable receivers, but for those starting from scratch, details of some companies that manufacture or market suitable receivers are given later in this book.

It will be appreciated that I am not in a position to test these receivers, so cannot provide recommendations as to suitability.

One good way of seeing receivers in action is to contact people in your local radio club, where it is most likely that someone will be able to help. In all probability a member will have a suitable receiver and may have already heard UoSAT-2, and be able to discuss your project with you. Radio clubs are excellent sources of help and advice. Here in Plymouth I met a number of people, one of whom was later able to provide expert help, solving a problem with my framestore. Another potentially useful source of help and advice is your local astronomical society. In Plymouth there are a number of such members who take an interest in this work.

If you already have a suitable receiver, you can go ahead and construct your dipole and tune in. Otherwise you may consider buying a general-purpose receiver, which includes other frequency bands, or you may choose to buy a specialist receiver – just for the UoSAT. You may decide to build a UoSAT (145.825 MHz) receiver yourself. Before doing so, first check what test equipment is required. A well-equipped radio amateur may already have an oscilloscope and other essential items, or be able to get easy access to them.

My original kit for building a UoSAT receiver was constructed by electronics trainees at the then Plymouth Information Technology Centre (ITeC).

The benefit of buying a dedicated receiver is that it should be optimized for UoSAT-2, and probably fitted with extra facilities such as a 'squelch' control. This allows adjustment of the level at which your (connected) tape recorder starts recording. Another feature of such receivers may be the provision of TTL signals; these are signals which are electrically modified (conditioned) to be fed directly into your computer's serial interface for program data processing.

With the type of system just described, you will *not* need to be present when waiting to receive UoSAT-2 data. By carefully monitoring your tape recorder, you can quickly see how much data is recorded during a 24-hour period. I usually set my squelch control to ignore lower passes, thus avoiding noisy data, and only recorded passes of good elevation – perhaps those over 30°. This could provide several minutes of quite good data which I could quickly process.

Examples are shown here of the various types of telemetry that I recorded and processed in the past. UoSAT-2 (and several other amateur radio satellites) remains a satellite experimenter's dream. And that is just the start! After UoSAT-2 you may wish to progress with the other UoSATs.

Data analysis

A dedicated UoSAT 2 receiver, such as the Astrid model, provides output suitable for direct input to a computer's serial input port. Some computers may not have a fully operational port so you would need to check this from your computer's manual, rather than risk disappointment when trying to connect the two. Using an Amstrad CPC6128, I had to purchase the serial interface which connects to the computer and allows the cable from the UoSAT receiver to be connected directly.

Conclusions

The above equipment serves to provide several projects. You can work out whether you wish to build some of it yourself. Later projects become more complicated, costly and involved.

Using the system described, you can monitor telemetry and learn to recognize various types of data being transmitted, simply by listening carefully. You can recognize 'real-time' data, 'whole-orbit' data, and ascii (text) data.

At this stage you would have been using a computer to analyse the data – unless you are seriously adept with a calculator! Various programs are available to extract different types of measurements from the telemetry, and you can appreciate the events that have taken place to enable you to get this far. Perhaps you have used an old BBC computer – or a 'PC'. Neither of these machines were available during the 1970s. At that time, a 40 Mb disk drive was some three feet tall and could only be carried with difficulty!

My first PC – bought a few years ago – had a 42 Mb drive, weighing just a few hundred grammes, and is about 10 cm diameter. I shall be taking a more detailed look at computers later.

Meanwhile, get some practice at data processing and aim to produce some scientific results from your data. The University of Surrey (address in Appendix 1) will be able to provide the latest satellite situation report on UoSAT-2 and will be interested to see your results.

7 Weather satellite equipment

Having covered the nature of weather satellites (often abbreviated to WXSATs), we now look at the specifications of the hardware required to receive their signals. The following chapter covers telemetry decoding (the production of a picture from the audio signal).

The UoSAT project (detailed in the previous chapter) can provide valuable experience at minimal cost, but is not a pre-requisite.

Complete systems

A complete system may be defined as one which receives and displays METEOSAT pictures alone, one covering reception of the polar orbiters, or one covering both groups.

Finances permitting, you can buy a complete system, and be reasonably sure that everything will work when connected together. Such systems are available from some suppliers – see Appendix 1. Price tags reflect development costs, and must be offset against the time otherwise required to design and build your own system.

Specifications of complete systems may vary widely. If such a system is your choice, read this chapter and the next – before purchase – to help identify the pitfalls and features of which you should be aware. The system must be designed to ensure that everything will be compatible.

Decisions, decisions!

Consider your aims during the next year. You might decide at a later date to expand your 'NOAA' system to include METEOSAT; conversely, you might wish to expand a METEOSAT system to include the polar orbiters. Will you go into computers? Do you prefer to do it yourself?

There are numerous ways of setting up an efficient ground station; become aware of future expansion options right at the start. You may decide to build some parts yourself, and buy the more difficult items – that will save time. To give you some idea of time-scales, you can expect to spend tens of hours constructing and checking out a simple receiver; if you build a frame-store – whether from a complete kit, or by collecting diagrams and components – you can expect to spend three, maybe six months, assuming that several hours per week are spent in construction. This book does not cover framestore construction or testing.

Computer?

Because of the availability of competitively priced, powerful computers, many go along this route, perhaps already having a suitable machine at home. A variety of software is available for several types of computers, not just the PC (Personal Computer – one which is IBM compatible). Satellite tracking programs are advertised widely and can be obtained through 'shareware' and 'public domain' outlets – see Appendix 1.

One advantage of using a computer for picture production is the ability to manipulate images after collection. The effectiveness of this depends on the computer, and software facilities. Give thought to other uses for a computer permanently in your ground station. Chapter 8 on telemetry decoding includes advice on computer specifications.

METEOSAT systems

A dedicated METEOSAT system consists of a dish/yagi, pre-amplifier, possible down-converter, receiver, framestore/computer, and display. Avoid buying each item from a different supplier, and this should minimize your chances of having 'post-purchase' problems. One manufacturer's pre-amplifier may not operate properly with another supplier's down-converter! The receiver may not be compatible either. Cost systems on the basis of literature specifications, perhaps asking for a discount if you are buying several items at once.

It really is not just a case of matching impedances between hardware! The output of the pre-amplifier may be far too high for the down-converter, for instance. One supplier could blame other items even if their own doesn't appear to work properly. Without extensive testing, you cannot be sure where the fault lies – so avoid the situation.

Both polar and METEOSAT systems can use the same 137.50 MHz receiver, and the same decoder – framestore or computer; this can minimize costs. It does, of course, mean that if you use your 137.50 MHz receiver for both systems, then you can only use one system at a time. You may want to get

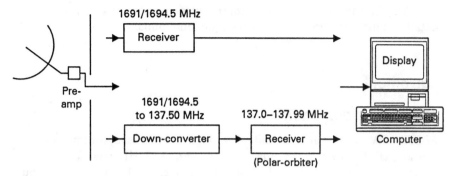

Figure 7.1 *METEOSAT systems*

a second receiver, in order to watch METEOSAT images while monitoring the polar satellites.

The whole process

METEOSAT transmits WEFAX on 1691 MHz (and 1694.5 MHz), so a yagi or parabolic dish must be used to collect the transmitted signal. For optimum results, a low-noise pre-amplifier is recommended, but may not be essential – particularly if you are experimenting. The amplified 1691 MHz signal is fed to the next stage – either down-conversion or direct reception. Some systems use a down-converter to transform the 1691 MHz signal down to 137.50 MHz, then feed it to a suitable weather satellite receiver. Alternatively, the 1691 MHz signal can be fed directly to a 1691 MHz receiver. This allows independent METEOSAT reception, but may be a more expensive option.

- *Option 1.* The down-converter reduces the frequency of the signal to 137.50 MHz. This lower frequency does not suffer excessive attenuation when fed indoors, through cabling, and into the 137 MHz weather satellite receiver. Suitable converters currently cost around £180. For efficient operation, a separate METEOSAT connection should be available on the receiver, to avoid having to swap between polar and METEOSAT cables.
- *Option 2.* If a direct 1691 MHz receiver is to be used, a low-noise pre-amplifier is essential. The signal can then be fed via suitable cable to the receiver indoors. I use this system now, having upgraded from the previous option.

The receiver – either direct METEOSAT or polar orbiter – extracts video information and produces an audio signal, which is an accurate reproduction of the original 2400 Hz modulated sub-carrier. Decoding this audio signal, extracting the modulation information and converting it to a grey-scale picture, is the job of the framestore or computer.

Having had an overview of the whole process – we now look more closely at detailed hardware descriptions of the individual units.

Equipment

Connectors and cables

METEOSAT signals are transmitted at too high a frequency for transmission through UHF (or similar) connectors. You can use UHF cable for carrying polar orbiter signals in the VHF (very high frequency – 137 MHz) band – but not for METEOSAT (at SHF – super high frequency – 1690 MHz). For this, use BNC, F or N-type connectors only.

Dishes and yagis

For METEOSAT reception, you can use a yagi or dish. Yagis are usually fitted with about 40 elements, and have a lower environmental impact than dishes, but with slightly lower gain, i.e. the output signal may be slightly lower.

METEOSAT dishes normally have a diameter of one metre, though a larger one is specified by EUMETSAT for interference-free WEFAX reception. The reason for this concerns the regular testing of other METEOSAT craft positioned near the nominal longitude of the currently operating METEOSAT. At such times as the second satellite is operating, secondary (WEFAX) users using small dishes will receive signals from both satellites, causing interference and leading to picture degradation. Dishes of 1.6 m and above can be used for receiving the digital data stream (PDUS) – when you have other hardware! Signal strength for WEFAX is considerably improved with a larger dish.

D-I-Y dish

METEOSAT reception can tolerate surprisingly large dish 'figure' errors! It is possible to construct a dish from chicken wire, having gaps up to 20 mm – though a narrower mesh looks better! Mesh can be fitted to curved strips of metal, attached to a central boss. I constructed my first dish in this manner, and was surprised at its effectiveness. The metal strips should form a parabolic shape because the dish is focusing radio energy, from a distant satellite, on the dipole/horn; the centre is slightly deeper than the surface of a sphere.

The cost of chicken wire mesh is nominal – just a few pounds. You need to make, or buy, a suitable feed. Horn-type feeds or a simple dipole cut to 86 mm (half-wave – see later) total length, can be used. I bought my own dish feed, a dipole of very simple design.

Figure 7.2 *My first dish*

A signal strength meter is required to align the system – I used one already fitted to my 137 MHz receiver. By adjusting the position and orientation of the dipole, in relation to the dish, a position of maximum signal can be achieved. For METEOSAT (currently MOP-2 positioned over Greenwich), the dipole will be horizontal. Later, re-adjustment of the dish sections, if so constructed, can be made for optimum performance.

Pre-amplifiers

The use of a suitable low-noise pre-amplifier, attached directly to the output connector of the dish, is recommended. As with the polar orbiters, it is possible that you may avoid using a pre-amplifier; try the system before spending more money. In my early days with METEOSAT reception I did not use one, consequently noting some occasional noise on the pictures.

Starting with a system in prime condition should enable you to produce the best pictures. The passage of time causes some of the components to age, producing a reduction in efficiency. Water may slowly leak into cable connections, if they are not sealed correctly. The combination of individual reductions in signal strength inevitably reduces the signal-to-noise ratio, and so may degrade the picture. This is one reason for beginning with significantly better than 'the minimum'.

Because of the high cost of pre-amplifier components used near 2 GHz, a good quality pre-amplifier costs considerably more than one designed for 137 MHz.

METEOSAT pre-amplifier kits may sometimes be obtained – see list of suppliers (Appendix 1). Construction of microwave devices should not be undertaken unless you have considerable experience in this field.

Down-converters

Down-converters are normally supplied in a small box. An input for the 1691/1694.5 MHz signal and an output suitable for the 137.50 MHz signal should be provided. The latter is then fed indoors to the weather satellite receiver.

Modern down-converters contain low-noise devices such as gallium arsenide field effect transistors (GASFETS). These may provide enough gain to avoid additional use of a pre-amplifier. Current prices for such converters vary from £150 to £180.

METEOSAT receivers

Recent developments have produced new, low-noise components, previously available at prices only for the professional market. METEOSAT receivers can take the 1691/1694.5 MHz, and extract the audio data to provide output for the next stage – the decoder – whether framestore or computer interface.

Direct receivers require a low-noise pre-amplifier to be fitted directly to the dish or yagi; the result should be a virtually noise-free image. Whatever system is used, output is an audio signal, ready to feed to the decoder.

The remainder of the hardware consists of the decoder (whether framestore or computer) and is essentially the same for both METEOSAT and the polar orbiters. Decoders are covered in the next chapter.

Primary data from METEOSAT

The hardware

A complete system comprises a dish 1.8 m diameter or larger, fitted with a feed horn, preferably adjustable, feeding a very low-noise pre-amplifier. The successful reception of primary data requires adherence to the specifications drawn up by the PDUS designers, and originally required a dish in excess of 2 m to achieve the required signal-to-noise ratio. It is a tribute to the advancement of electronics design that ultra-low-noise pre-amplifiers now permit the use of dishes of a size smaller than 2 m.

The 1.8 m dish and pre-amplifier allows some twenty metres of cable to carry the 1694.5 MHz signal indoors, into the specially designed receiver. This unit has little in common with a standard METEOSAT WEFAX unit, being of a digital nature, with far wider bandwidth. Basic receiver controls are set up during the installation.

A dish requires some form of mounting and, for quick alignment, can be set up connected to a WEFAX receiver, in order to point the dish reasonably accurately at the satellite. The continuous WEFAX audio tones from METEOSAT allow quick alignment. A hardware expansion card is fitted into the computer to receive data from the receiver for digital processing.

Primary data involves a considerably higher rate of flow of image information than does WEFAX, so as expected, the imagery is not only higher quality, there is more of it.

My first impression of PDUS images was their clarity and resolution. Zooming in to an AV image (whole disc, visible image) allowed the examination of the image to its maximum detail. Once per day, this 2.5 km (at the sub-satellite point) resolution image reveals the true nature of the landscape at places all over this hemisphere of the globe. The clearest imagery of Africa allowed me to study the drought-stricken regions of the continent, to an extent not fully appreciated by WEFAX users, even though those images are of very good quality.

The future

As hardware prices continue to fall, one can expect an increase in the numbers of people using primary data systems. I do not expect to see a huge increase, on the grounds of current dish requirements, and the cost of de-encryption units. Few people can install large dishes without bolting them down properly, and to do that requires planning permission.

I expect that if ultra-low-noise pre amplifiers are developed to the extent that dishes of about 1 m diameter can be used, we may see the demand grow.

Meanwhile, there is a lot to be said for installing METEOSAT computer-based WEFAX equipment for the time being. By the time that pre-amplifiers have progressed further (perhaps 12 to 24 months), the only additional items needed for the upgrade could be the pre-amplifier and the receiver/hard card.

Polar-orbiter systems

Complete systems are available for reception of a.p.t. from polar orbiters, and these consist of a VHF antenna, a pre-amplifier (where considered necessary), cable into your 'operations room', and a dedicated weather satellite receiver, framestore or computer and display.

Antennae

Firstly, the principles of antenna use, and how your choice could relate to your proposed use of the signal, are similar to those with UoSAT-2 equipment. You

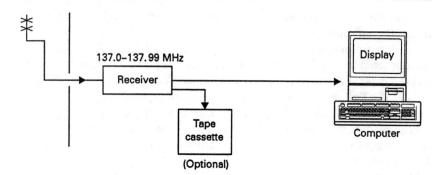

Figure 7.3 *Polar orbiter system*

can use almost any tuned antenna to 'receive' weather satellite signals, but the resulting signal quality depends on the antenna.

Monitoring such satellites can be done on a shoe-string – almost literally! The strength of the VHF signal (average 5 watts) from weather satellites is enough to enable you to tune into them using a discone, simple dipole, crossed dipole, or even an odd length of wire (called a random-length) pushed into the back of your receiver!

When my first receiver was completed, I stuck a short length of wire into the antenna socket and immediately heard a NOAA satellite passing. That wire would not receive a good signal for more than a minute or two, because of its reception characteristics. The antenna used should be chosen on the basis of your proposed application for the final signal. For monitoring satellites, a discone or other reasonable antenna will suffice. Do not, however, expect to produce superb pictures using a discone!

Simple dipole

The length of a dipole can be estimated:

wavelength = speed/frequency
$$\simeq (300 \text{ million m/sec})/(137\,500\,000\,\text{Hz}) = 2.18\,\text{m}$$

This is the true wavelength, but a rather unwieldy length. It is easier to cut a dipole to a fraction of this dimension; half-wavelength = 1.09 m or quarter wavelength = 0.55 m.

Allow for the change of velocity in the dipole; it is about 5% slower than in free space:

actual total length across half-wave dipole = 1.145 m

This is a convenient length, corresponding to half a complete wavelength – so we call it a half-wave dipole. There is no mystery about using a 'half-wave' instead of a 'full-wave' dipole. The generation of 'resonant' signals in a length of wire (the antenna) is a complex subject. For a specific wavelength, multiples and sub-multiples (we use the term harmonics) of this wavelength are generated. Consequently, we can use much shorter lengths of wire than seem suitable at first sight, using these calculations.

A properly cut dipole can receive a.p.t. signals, but the received strength will be about one-half of that available. The original signal is right-circularly polarized, so for proper reception, a properly phased crossed dipole is required.

Crossed dipole

Such a dipole, cut for the 137 MHz band and correctly phased for right-hand circular polarization, is the optimum type. All but the Chinese FENGYUN satellites use this type of polarized signal; perhaps someone wired up FENGYUN wrongly? For notes on the correct phasing of a crossed dipole see the 'antenna phasing' section in Chapter 6.

You may decide to purchase a commercial crossed dipole. If so, do check that the dipole lengths have been cut correctly! There have been instances of such antennae having dipoles that were too long or short for the band.

Connectors – N-, BNC, UHF-, F-type

Having constructed or bought your crossed dipole, you need a suitable connector fitted to deliver the signal. For many years, the most common connector was the N-type. These are, theoretically, the best for connection between transmission systems operating above 200 MHz, and are suitable for use beyond 2000 MHz.

For our purposes, N-type connectors are overkill! They are designed for high operating voltages, rather higher frequencies than 137 MHz, and can be expensive. In my view, the aggravation involved in trying to fit an N-type connector to cable seems completely unnecessary. One alternative – the BNC-type connector – is often used in small diameter cable systems.

If you choose to buy an N-type connector, try to obtain one which includes a sheet giving construction dimensions. There are various types of N plug, all having small differences, such as the internal pin diameter. Without proper instructions, you may not be able to assemble them correctly, unless you are an experienced electrician. In the worst case, the plug may come apart after a few weeks of manipulation.

If there is an N-type connector already fitted, you can connect one N-type plug to the down-lead cable, or even use an N-to-UHF connector, all easily available from amateur radio rallies, and some electrical shops. UHF plugs are simple in the extreme; their limitation is their unsuitability for frequencies above about 200 MHz. Don't use them for your wide-band scanner!

Another antenna connector increasingly used is the F-type connector. This is my preference, and they are used extensively on satellite television cable. Whatever combination of plugs and sockets used, take your time in assembly, and test the system at ground level. Without realizing the necessity for this, I paid for a roof antenna to be mounted on the chimney, three floors up, before I even tested it. A manufacturing fault in the phasing harness had to remain for two years before I could pay for the antenna to be removed and a fully ground-tested one installed!

Low-elevation passes

Details of VHF antennae suitable for weather satellites are covered in later chapters. To summarize – a crossed dipole (right-circular polarization) is the preferred antenna, though more directional helixes have been used successfully. If you wish to receive telemetry from NOAA (or METEOR) satellites near the horizon – i.e. when at their maximum distance from your station, you can experiment by mounting a normal crossed dipole, pointing northwards or southwards at lower elevations. I had some success with the higher METEORs, as they passed over the North Pole, although greater terrestrial interference was also experienced. The optimum angle for good reception should be around 30° above the horizon.

Cable

Receivers used with weather satellite systems normally specify an antenna of 50 or 75 ohm impedance. The receiver largely determines the antenna impedance required to match. Most dedicated weather satellite receivers specify an input impedance of 50 ohms, so for optimum results you should be getting 50 ohm cable and connectors. In practice, after operating your system successfully for some weeks, you may start experimenting and find that even 75 ohm cable from a UoSAT antenna, produces a fair signal!

Impedances are specified, but may not be too significant. I once spoke to an engineer who told me that his UoSAT receiver (which he had designed commercially) had a 'nominal' (or average) impedance of 50 ohms but was happy with almost anything. I feed it with a second weather satellite antenna, fitted with 75 ohm cable.

Check the receiver's requirements, and select cable accordingly – there are several varieties available. Cable has two important characteristics: impedance and attenuation loss, the latter measured in dB per 100 m. The cable carries the signal from the antenna into your receiver, so there must be some losses. An attenuation of 3 dB means that half of your signal is lost – it is converted to a minute amount of heat during transport.

The best cable should have the lowest loss at 137 MHz. I used H100, which was highly recommended, and is good. As with UoSAT-2 you should aim to use no more than 20 metres of cable to feed the signal to your receiver. Thicker cables are more expensive, but usually have a thicker central copper core. You may not need to go to such extremes if you only wish to monitor signals.

UR67 and UR43 cables are good, the thicker being perhaps more appropriate for cable for the higher frequencies used by METEOSAT. I use H100 (a high-quality cable) for the 18 metres run from the roof down to my room. I also tried a different antenna which came supplied with 20 metres of narrow 50-ohm cable, and found this combination to be perfect. For moderate (less than 20 m) cable runs, the highest-quality cable may not be essential.

Pre-amplifier required?

Before going to the expense of buying (or constructing) a pre-amplifier, test your antenna (and cable) at ground level by monitoring a high pass of a weather satellite.

Using a proper tracking program you can test reception characteristics within a few hours. If you find that you can hear the signal within a short time of the satellite rising above your horizon, you may not need a pre-amplifier. Allow for the reduced horizon at ground level. High-elevation passes should provide a perfect, clear signal for several minutes. If not, check cabling before assuming that you need a pre-amplifier. Good-quality cable, with properly fitted connectors, should be satisfactory for runs up to 20 metres.

Pre-amplifier specifications

If you really do need to buy a pre-amplifier, you should buy one that provides just enough gain to compensate for cable losses. This should be within about 10 dB, so avoid pre-amplifiers with huge gain. The noise figure of a pre-amplifier is its most important characteristic – the lower the better. Look for no more than 1 dB. Power, supplied between 10 and 14 volts, can normally be provided along the down-lead cable. Out-of-band selectivity refers to the efficiency with which the amplifier avoids amplifying signals outside the band. The pre-amplifier should only amplify signals in the 137 MHz band; gain should fall off rapidly outside the 136 and 138 MHz limits.

Reception limits

Using a roof-mounted antenna, from most localities you should receive signals from low-elevation passes of all the weather satellites. Experience soon reveals just how low you are able to monitor satellites around your horizon.

Weather satellite receivers

Many receivers on the market are 'designed for weather satellites', but it is necessary to select your receiver with care, whether in kit form or built and tested. My own enquiries and experience have shown what a minefield this subject is!

Receivers can be bought as part of a complete system, in which case questions must be asked about the circumstances in which the receiver will be operating. Read this section before buying! The topics of crystals, synthesized frequencies, intermediate-frequency bandwidth, filtering and paging interference are covered next.

Crystals or synthesized receivers

Frequencies within the receiver itself can be generated by individual crystals, or synthesized by chips. Opinions concerning the best methods of synthesis vary between receiver manufacturers; some make both types. The benefit of synthesized receivers is the large number of frequencies that can be synthesized within one receiver. One of my scanners can tune to 200 channels between 136.00 and 137.99 MHz; I occasionally let it scan every frequency in these bands, in order to monitor any satellite transmission, whether expected or not.

Crystal-operated receivers use individual crystal oscillators to generate each frequency, so a weather satellite receiver is normally fitted with a limited number of crystals – perhaps those for 137.30, 137.40, 137.50, 137.62 and 137.85 MHz. A problem can arise when new satellites are launched, operating on non-familiar frequencies. The Chinese satellites FENGYUN transmitted on 137.04 and 137.80 MHz, for example. Manufacturers should be able to supply customers with suitable crystals.

These receivers may or may not have a scan facility. The required frequency is selected by pressing the appropriate button. The benefit of using crystals is undoubtedly the low noise figure. Signal purity is therefore higher with this type of receiver.

Sensitivity

Using the latest components, receivers have been designed to be exceedingly sensitive. Specifications still vary between systems; a good figure to expect is 0.2 or 0.3 μV. Such specifications may vary, even for the same system, depending on how the manufacturer measures the figure. The figure may also vary according to the quality of the measuring equipment!

Normal measurements may be quoted as follows:

0.3 μV p.d. for 12 dB SINAD at 10 kHz deviation

These figures describe the sensitivity over a specified bandwidth – i.e. a signal as low as 0.3 μV was detected, using a specified bandwidth. Another method describes the signal amplitude required to silence (quieten) the 'noise' that you hear in the absence of signal. This may be quoted as:

0.35 μV p.d. for 20 dB quieting

Intermediate frequency (i.f.) bandwidth

Remember the nature of the original signal. The satellite scans the Earth; the information from each line of the picture, amplitude modulates a 2400 Hz carrier. The final a.p.t. information results in a band of frequencies between about 800 Hz and 4000 Hz, representing the scene's brightness levels, ranging from black to white.

This band of frequencies, superimposed on the 2400 Hz carrier, modulates the main 137.5 MHz carrier, so the actual picture information is represented by a small portion of the 137.5 MHz signal. The first step in signal decoding (extracting the information) involves converting the signal to a much lower frequency, but one still containing the information. This lower frequency is called the intermediate frequency.

Receiver circuitry design involves the careful selection of filters to remove all unwanted signals generated internally and externally during the extraction of image data.

The r.f. signal has a bandwidth of about 30 kHz, so if the satellite was not moving, the receiver could have a similar bandwidth and work effectively. This bandwidth is unusually wide. Receivers having smaller bandwidths, such as narrow-band frequency modulation (NBFM), often the case with a general-purpose receiver, will simply not pass the whole composite signal through the electronics. Bandwidth is therefore a critical factor in selecting a good receiver.

It is entirely feasible to receive satellite signals on general-purpose receivers – I use my Tandy PRO-2004 for such purposes, but many similar receivers will do. Listening to weather satellites using this receiver (which uses NBFM for this band) one can hear what appears to be a clear signal. Unfortunately over half the picture information is missing! So if your only aim is to monitor satellites, you do not need to spend out on a dedicated weather satellite receiver. For those wanting to decode pictures we must look further.

Doppler effects

This 30 kHz bandwidth is not the whole story. The satellite is moving relative to the stationary observer (and antenna); this affects the received signal. When the satellite approaches the observer, the frequency appears to be raised – the well-known Doppler effect, heard at every railway station as trains approach. The apparent frequency may be raised by up to 20 kHz, then lowered by a similar amount as the satellite recedes. The actual bandwidth can be up to some 50 kHz. If less than this, the receiver may lose some of the signal content.

An acceptable bandwidth is about 50 kHz, and the best receivers are specially designed for such reception.

Filters

Circuitry used in such receivers includes combinations of filters, and can present a bewildering list of technical specifications. Active, low-pass, high-pass filters may all be listed. The real question is – will they give you a good signal in the presence of unwanted noise? Specifications do not completely answer this question.

Experimenters can try new filter combinations in part of the receiver to try to minimize local problems. They may be successful, but if not, further modifications may be made until the best one is found.

Studying advertised receiver specifications may not necessarily guarantee success. One answer is to experiment with your own receiver circuitry, or, if you prefer, buy a receiver, with an assurance from the supplier that if it proves unusable in your location, you will be able to return it.

Paging interference

The national allocation of frequencies to paging devices has severely encroached on the 137 MHz weather satellite band. Satellites operating here transmit about 5 watts. Paging frequencies were allocated as low as 137.9 MHz, at which frequencies they may output hundreds of watts! It is hardly surprising that all over Britain, we initially suffered severe interference from these transmissions.

It has proved possible to modify some receivers to minimize the effects of pager signals, though some manufacturers appear to have done little to minimize or prevent such interference to their products! Be careful when selecting kit, or buying a completed receiver.

I received many letters from *Short Wave Magazine* correspondents who successfully modified their receivers and eliminated local interference – in every case it was a kit version that was modified.

Facilities

Whether you buy a kit or complete receiver, you need several facilities. Scan, direction, override, halt, volume, squelch, outputs – all may be found desirable.

If you choose a scanner, there should be a number of facilities:

- *Scan direction* – whether increasing or decreasing frequencies may be selected
- An *override* switch can lock out frequencies temporarily suffering interference, or force scanning regardless of the presence of signals
- A *halt* switch stops scanning, and is used to remain on a selected frequency, regardless of the absence of signal
- *Volume* control is for audio output

- A *squelch* control enables you to set the level at which the scanner locks on a signal. If this is set too high, you may miss low strength signals, and if it is set too low it stops on noise bursts
- There may be *output level* controls for setting the level to your computer's decoder. This should normally be set at 1 volt (peak-to-peak); there is usually a corresponding control on the computer interface. A second output, at a much lower level, may be provided for tape recording the audio signal
- On some scanners, there might be a *page* switch, used to select a specific band of frequencies for scanning. Bands are divided into groups, e.g. 136.00 to 136.99 MHz, often in steps of 10 kHz

If your receiver is a fixed frequency model, several of these options are not relevant.

Receivers for PDUS and h.r.p.t. systems

During the early 1990s there was little information available on PDUS and high-resolution picture telemetry (h.r.p.t.) receivers. A number of American experts built their own high-resolution and primary data user system (PDUS) for monitoring METEOSAT data, and h.r.p.t for monitoring the 1700 MHz band from the NOAA satellites.

A few companies and institutions have developed these systems to the extent that they now offer complete systems at costs almost approaching 'domestic' level prices. One or two universities and colleges offer systems for sale at prices in tens of thousands of pounds.

PDUS

A METEOSAT PDUS receiver is completely different from that of a WEFAX receiver. Considerably wider bandwidths are required, and other major design changes. A high-quality pre-amplifier is essential, as is a larger dish. A 1.8 m dish can suffice. A beginner starting on PDUS or h.r.p.t. is perhaps rather like a newly qualified driver wanting to buy a Rolls-Royce. May I suggest that you start at the beginning, then if you later want 'the best' or 'the latest', at least you will have had the valuable experience of WEFAX and a.p.t. The additional cost of a de-encryption unit (currently likely to sell at around £500) must also be included.

NOAA h.r.p.t.

Equipment suitable for the reception of h.r.p.t. is available, though again, it is in a much higher price bracket. You cannot use an omnidirectional antenna for 1690 MHz band transmissions from a moving satellite. A fully driven system is a necessity, in order to track the selected satellite. Minimum dish size is about 1.2 m, as specified in recent advertisements. PDUS also requires the highest-quality pre-amplifier, currently costing over £200. Receivers specially designed for h.r.p.t. currently cost about £700, and a suitable decoder adds about £400, assuming the computer-compatible unit was purchased.

For the time being, high-resolution pictures may remain above hobbyist price ranges, but no doubt time will play an important role in price reduction. If you seriously contemplate purchasing an advanced system, at least appreciate that only a few years ago, such hardware was the exclusive field of the professional.

Back to a.p.t.: the output signal

Whatever type of receiver you instal for satellite reception, the final output is an audio signal containing picture information. The next stage in the process is the decoding of that signal. The choice of decoders will be discussed next.

8 Decoding the image

Having set up an antenna, feed and receiver, a clear signal from a weather satellite should be heard from the speaker–when one rises above your horizon. An output from the receiver provides an adjustable signal which can be fed into a decoder; this should extract the original picture data from the signal. It can only decode efficiently if the receiver is properly designed to extract the full range of frequencies from the original signal transmitted by the satellite. If the receiver is unsuitable for weather satellites, the decoder cannot perform this function properly. If all is optimized, the decoder should provide a good-quality video signal which can display a picture truly representing the original view from the satellite (excluding colour).

Decoders vary in quality and cost, depending on the facilities provided, and the electronic skills of the manufacturer. This chapter looks at various options to assist you in making your decision.

The framestore

By the middle of the 1980s, many radio hams and electronic hobbyists were decoding weather satellite pictures. This followed the publication of an article on building a framestore, designed by Matjaz Vidmar (call sign YU3UMV). This became the most widely known framestore, often just called the YU3UMV framestore. Since the original article was published (see *VHF Communications*, April 1982 and January 1983), many modifications have been made to the circuit design in order to expand the capabilities of the original unit.

I partly built my own framestore, but had the all-important boards professionally constructed. The project took over three months to complete, and the final checking (de-bugging) was done by a member of Plymouth Radio Club, Joe Jones.

Since those days, the increasing popularity of this hobby has brought commercial interests into the field, and a number of framestores or similar hardware have become available. Kits are still sold to the advanced electronics hobbyist. This general description of the operation of the YU3UMV framestore should enable you to identify features that you might want on a commercially manufactured product. This option can also be compared with computer-related facilities.

It is well outside the scope of this book to provide circuitry and construction details for the YU3UMV framestore, but they may still be obtainable from members of specialist clubs – see Appendix 1. The unit contains individual circuits to provide synchronizing signals and video output, and can decode all weather satellite formats – NOAA, METEOR, METEOSAT and OKEAN. It can display the full satellite resolution, that is, can show all the detail available from a.p.t. transmissions, and has a facility to invert the picture being received, in order to present southbound satellite pictures correctly, as well as northbound ones.

Modifications

Les Currington, G8LOK, designed a significant modification to the original circuitry – a reference 2400 Hz tone generator, allowing perfectly aligned imagery to be decoded from Russian METEORS. A further modification provided a facility to adjust the dynamic range (span) of the analogue-to-digital convertor (ADC). This unit converts the continuous (analogue) waveform into digital form (binary).

Visible-format images have a wide dynamic range (remember picture information is represented by a wide voltage change) and can provide the maximum number (normally 64) of grey levels of which the electronics are capable. Infra-red images have considerably smaller dynamic range, and without 'real-time' adjustment, the framestore may only provide a few grey levels – resulting in a dull, featureless image. By adding suitable circuitry – an adjustable ADC – to the framestore, this problem is overcome and the capabilities of the system are considerably enhanced. The additional circuit provides 'black' and 'white' level controls, allowing real-time adjustments to be made for optimum results.

Other controls include a synchronizing switch to cause the picture to start at the beginning of a frame or section. This can be used to select either infra-red or visible portions of a NOAA transmission; the circuit depends on the different tones transmitted before each section – as explained in Chapter 4.

In addition, a 'left–right' biased switch can be used to shift the image sideways to obtain a synchronized picture. Later modifications (published in various magazines such as the quarterly *Journal of the Remote Imaging Group*) allowed the storage of more picture data, particularly from METEOSAT. Further features were quickly overtaken by computer developments.

Framestore kits

Having read a description of the main facilities of a framestore, those who wish to construct one should be aware of the major effort required. Not only is a good knowledge of electronics required, but also a comprehensive set of test equipment, or at least access to such.

A study of electronic hobbyist magazines such as *Maplins* and *Cirkit*, shows that a number of kits have been available for the construction of framestores and decoders. These have been successfully built by a number of people, including *Short Wave Magazine* correspondents. Articles usually emphasize the construction effort and testing gear required. Sometimes there is the added benefit of a fault-finding service, though the potential costs of this labour-intensive work must be appreciated. At the end of the day perhaps only the most dedicated electronics constructor would wish to attempt these complex projects.

Artificial colour

The addition of artificial colour to an image can be very impressive. You can achieve realistic 'colour images', but do not forget that the satellite does not 'see' in colour. The received signal does not contain any colour information. All that the hardware – framestore or computer – can do is to substitute, for example, blues for dark areas (sky or sea), green for those rather lighter (usually land), and grey–white shades for the lightest (cloud). Each grey level is associated with an intensity level – usually either 64 or 256 individual levels; false colouring simply means replacing intensity levels with specific colours. Those chosen are near enough realistic to be pleasing to the eye. It can also be useful to add a range of red hues to infra-red imagery, to represent temperature differences.

Computers

An overview for the satellite monitor

Rapid progress made by computer manufacturers lies in the research done in electronics. During the 1950s and 1960s huge strides were made in reducing the physical size of computer memory – initially using valves, then transistors, microprocessors (chips), and now integrated circuits. Impetus was given to computer technology development by the requirements of the American manned space programme.

A medium-sized (mini) computer was capable of processing the most complex scientific data for the British space programme, and was also used to do jobs that would one day revolutionize the office.

By the early 1980s, several incompatible small computers were developed; the BBC computer, Commodore Vic and a fleet of small 'cult' machines. Programs written on one machine would not run on another.

IBM entered the field of table-top computer manufacturing in the 1980s, producing the IBM Personal Computer (PC). Its marketing muscle was such that its machines would always be adopted as the 'standard'. The market for IBM software grew considerably, so programs were developed by major software houses. Amateurs also wrote software.

We can now buy a decoder which slots into a vacant expansion slot in the computer. It will include software which, after correct installation, will produce a weather satellite picture from the audio signal fed into the connector on the external edge of the card. Before detailing the options, a look at basic computer specifications should be of use.

Computer requirements

I am often asked 'What is the best machine for me to buy?' Before purchasing a machine, you need to spend time considering the jobs for which you expect to use the computer.

Some possibilities:
Business, correspondence, documentation, regular financial monitoring, job planning, record maintenance, games, hobbies.

There are many applications for computers in the fields of radio amateur use. Glance through the different sections of amateur radio magazines and you will see reports from people, monitoring all parts of the radio spectrum. If you are considering the purchase of a computer specifically for work in this field, for example satellite tracking, RTTY and picture decoding, it is almost essential to have some technical knowledge before purchase. The following guide should be of interest.

TOPICS

- Size
- Compatibility
- Co-processors
- Floppies
- Speed
- Monitors
- Parts

- Expansion
- Processors
- RAM
- Storage
- Maintenance
- Video cards
- Cache

Footprints

Computers come in different shapes, commonly referred to by their 'footprint' (desk coverage). Small units may be described as 'slimline', then, in increasing size, come the 'desk-top' (the most common type), the 'half-tower', and the 'full tower'.

Check the case construction – you want all metal, not plastic! Metal casing should help to minimize unwanted electrical interference coming from the computer – of critical importance when you are using it near radio receivers!

One reason for different-sized cases is to allow for different expansion requirements. A slimline case cannot easily accept further hard drives or even new boards for its internal expansion slots. At the other extreme, a tower should have space for a variety of expansion possibilities.

Expansion slots

Inside each computer is a motherboard containing the actual processor: a 33 MHz 386DX or perhaps a 25 MHz 486SX, and other critical components such as RAM – see later. Fitted to this board are a number of slots, designed to allow for future connection of peripheral devices, including accelerators or even a FAX, normally supplied on cards. These come in two sizes – half and full cards.

My own half-tower has no more expansion slots. They now hold a CD-ROM, second disk drive, and cards for a.p.t. and PDUS – some of which were not anticipated, but were allowed for, at the time of purchase.

Compatibility

This is the ability to run 'IBM' software. Computers such as the Archimedes have their own (different) operating system, so cannot normally run programs designed for IBM and clone (copy) machines. There is a type of software available – called an emulator – which allows some IBM programs to run on some non-compatible machines. Emulators achieve compatibility with varying degrees of success. Some years ago I used an emulator on my Nimbus computer, but it was prone to crashing; my next upgrade was an IBM clone.

Other terminology used to describe computers includes XT (extended technology) and AT (advanced technology). Avoid XTs. These were the earliest IBM machines of the PC type, and are virtually impossible to upgrade. Some are still available at very low prices.

Processor speed and type

The speed at which a computer 'runs' depends primarily on the operating speed of its main processing chip. The main types are the 80286, 80386 (DX) and 80486 (DX), the Pentium and some SX variations. Processors of the 80286 type usually run between 12 and 25 MHz. Faster speeds are available with 80386 and 80486 processors. Such speeds are not normally required for most data collection projects, but other 'routine' applications software will run more efficiently.

Some early applications software, used for image decoding, ran inefficiently because, with a slow processor speed, digital analysis techniques could not operate quickly enough. These compromises meant that picture quality was not good. Fortunately we are well past that era, but some cheap, second-hand 8086 computers are still available–leave well alone. If you wish to run 'Windows' applications, even a 386SX is at the bottom of the recommended machines list. The 386DX was the first microprocessor to be able to run more than one program simultaneously–to 'multitask'. This feature can be useful, but you have to be competent to install and use such facilities. The 386SX is a slimmed-down version of the DX chip, and operates slightly more slowly.

Cache

The 80486(DX) chip is similar to previous processors, but has an on-board co-processor (see later) and some cache memory. Cache is a special type of memory storage–it is extremely fast (much faster than conventional RAM), so can speed up the flow of instructions to the main processor. Without cache, the processor may have to wait for (slower) RAM to provide the next set of instructions. Although cache is not essential, most machines will include it; look for at least 64 kb cache–more is fine but may not affect the operating speed significantly.

Recent developments in chip technology include the Pentium (P5 or 586) processor. Specifications may include an operating speed of 66 MHz and higher, with other refinements. Experience suggests that although prices are high, and its capabilities far outside our requirements, prices must fall and we may soon consider it standard.

Maths co-processors

Co-processors speed up mathematically intensive calculations, so are of great benefit when manipulating numbers, but they are not essential. The 486DX and higher chips include an operating co-processor within the chip structure. The 286, 386SX and 486SX processors have no operating co-processor. I fitted one in my 386DX, which greatly speeds up predictions and tracking software.

RAM

Random-access memory (RAM) is the built-in memory available for programs and data. A disk operating system (DOS) normally uses up to 640 kb, but it has long been possible to add extra memory above this DOS limit, and some programs can now use this effectively. For Windows applications your machine needs a minimum of 4 Mb to run efficiently; most new computers have at least 4 Mb RAM fitted, and for improved performance with some software, you may specify more. Installing extra RAM generally costs £20 to £30 per megabyte. Increasing use is being made of extra RAM by recent imaging software, by holding animated images.

Floppy drives

These are the 5.25 in or 3.5 in drives usually mounted on the front of the machine, for loading in programs and data. There are two types (densities) for each disk size. Modern floppy disks store 1.2 Mb (for 5.25 in), or 1.44 Mb (for 3.5 in). This capability is called high density (HD). Lower-capacity disks are called double density (DD).

If you specify a drive, go for a 3.5 in – they are far more rigid. When buying new disks, remember to ensure that you buy the correct density (HD) and avoid purchasing what are called 'bulk' disks. Your drive should be able to cope with both densities. If buying second hand, check the type fitted.

To appreciate the significance of kb, it may help to know that a sheet of A4, filled with text, occupies about 2 kb of storage. You can hold the contents of an average book on one disk.

Storage space

The size of the hard disk(s) defines the storage space for programs and data. Until a few years ago, the minimum size considered workable was 42 Mb. After adding a few multi-megabyte programs, you may not have much left for data storage. Some applications produce large data files – my METEOSAT primary data system can produce individual files up to 30 Mb, so I fitted a second drive.

File compression programs compact data, without loss of integrity, before storage. They therefore improve storage efficiency. The problem with data compression is the resulting dramatic reduction in the speed of program operations. It is feasible to run tracking software from a compressed disk, but Windows applications almost grind to a halt!

You may decide to add a second hard drive at a later date; check your computer has a suitable expansion port and cable attachment for a second drive. Some 'small footprint' computers are too small for such expansion.

Disk access speed

Another characteristic of hard disks is their access speed. The time taken for the disk drive head to move to a specific location on the disk and write or read data has reduced dramatically. Some disks are now quoted as having access speeds as fast as 10–12 ms. For many applications, slower speeds of 20–30 ms are fast enough. These speeds are almost meaningless without regular disk maintenance.

Disk maintenance

After several months' use, disks contain an increasing number of fragmented files – those where data is spread over different areas on the disk. This is a consequence of the manner in which DOS works, and, over a period of time, can reduce the disk's effective performance. Using suitable software, you can re-combine (de-fragment) the file sections, and combine (make contiguous) empty spaces, enabling the system to write data to one large area – a fast process! Utilities such as 'defrag', now part of DOS software, can perform this task.

Another feature of DOS usage is the innocuous loss of small areas of the disk, which occurs during continual file storage and erasure. The loss is not physical – the storage area remains – but DOS can lose track of it. The problem is easy to correct; use the program 'chkdsk' (check disk), which is found in the DOS directory (version 6.2), already on your computer. The DOS manual or the built-in DOS help screens explain how to do this.

These facilities can increase the efficiency with which your computer operates.

Monitors, pixels and video cards

The standard monitor barely exists. The variations can be confusing to someone who simply wants to purchase a machine. Monitors are character-ized by type (colour or black-and-white), size of screen (14 in being 'standard'), and by the number and size of pixels (picture elements), across and down the screen. Monitor resolution depends on dot pitch size, normally quoted when you buy your computer. One former standard was 0.31 mm, but for best quality, look for 0.29 mm dot pitch.

For black-and-white monitors there is one single-colour beam illuminating each pixel, giving generally sharper images. Colour monitors require three single-colour beams, each having to focus on the same pixel to within 0.29 mm! Beam intensity must be specified, so the more pixels used to display a picture, the greater the memory requirement for the monitor control card.

Let's indulge in some revealing mathematics. Each pixel has an individual address, stored in the computer, and used by the video card. Enormous memory is needed to tell every pixel – which of many colours it should be.

A VGA monitor can display 800 pixels across, and 600 pixels down, so has a total of 480 000 pixels. Each needs information defining its brightness level and colour. One byte of memory represents eight individual bits, so can represent any of 256 (2 bits raised to the power of 8) different levels. This (800 × 600) display requires 480000 bytes, so a card driving the monitor must have at least this amount of memory fitted. Consequently, for this display quality, you must confirm that the card is properly populated with sufficient chips – 512 kb in this instance.

The card

Such cards need RAM ranging from 256 kb to 1 Mb, just to hold pixel information – don't be confused with program RAM which actually generates the picture. Sometimes the video card can be upgraded, perhaps by installing more memory chips. This extra memory then enables higher-resolution monitors to be used – or perhaps allows the monitor to display at its maximum capability.

Monitor standards

There are so many that I shall only cover the main types.

CGA

The colour graphics adapter (CGA) was one of the earliest types, introduced in 1981 as an IBM standard, having the most rudimentary graphics capabilities. The CGA screen has a few modes (choices of display resolutions), e.g. 320 by 200 pixels, each having one of four colours. Can you imagine trying to display high quality pictures in CGA? Don't! The only computers likely to use CGA displays are small, cheap portables and old, second-hand computers.

EGA

The extended or enhanced graphics adapter (EGA) originally supported 320 by 200 pixels, each having one of 16 colours. Later cards provided 16 out of a potential 64 colours at an increased resolution of 640 by 350 pixels – a great improvement. You may see computers with such displays, though they are all but obsolete. Modern software needs better than this for optimum results.

VGA

Video or versatile graphics array (VGA) arrived a few years ago, and provides a choice of 320 by 200 pixels with 256 colours, or 16 colours for 640 by 480 pixels, usually the full resolution. A choice of 256 colours includes only 64 grey levels (because each grey level consists of the three individual colours, which must have identical intensities). Some specialist graphics cards can provide 256 grey levels for some VGA monitors.

SVGA

The super VGA (SVGA) monitor is the current standard, though there have been further developments, e.g. extended graphics array (XGA). SVGA offers 800 by 600 pixels, each of which can be illuminated by any one of 256 colours, themselves selectable from a palette of 262 144. Of course this does depend on the software offering such choices.

Interlacing

Most desk-top computers come with SVGA monitors and the remaining variable is the 'interlaced/non-interlaced' type. With so many pixels to display, the monitor handles vast amounts of data quickly. It is expensive to design monitors to do this properly at the highest resolution, so 'tricks' are used!

An interlaced monitor scans half of the screen (alternate lines) in one pass; it then fills in the remaining (interlaced) lines on the next pass. Using this method, a monitor can show 1024 by 768 pixels. The price of this 'trick' is that the picture may appear to flicker because the eye is not fully deceived. Non-interlaced monitors display the full screen in every scan, but such monitors are expensive. For hobbyist applications, there seems little benefit in spending extra on a non-interlaced monitor.

Operating system

Every computer comes with a disk operating system (DOS) already installed. New versions of DOS come out occasionally, providing new facilities. Some people prefer to avoid DOS. The most well-known alternative is the 'icon' system called *Windows*. This uses the principle that pictures (icons) are more easily understood and remembered. Each system has its benefits and advantages. If you study the available facilities, you might be surprised at what you find you can do. Utility programs for decoding WEFAX, CW, RTTY etc., are usually DOS based, so Windows expertise may be kept for other applications!

An advanced operating system called *Warp* is the latest version of IBM's OS/2 software. This requires a machine with at least 4Mb RAM and oodles of hard disk space, but offers an effective environment in which a number of jobs can run together – or multi-task. In my view, the main advantage of Warp is this ability to simulate separate machines, rather more safely than Windows. Although Windows allows jobs to (nominally) multi-task, in reality, if one crashes the system, a re-boot (the simultaneous pressing of control-alt-del) is normally necessary, thus losing all the other jobs.

Ports

Communications with the outside world require a telephone. This converts human speech to audio signals, carried by the telephone network. Your computer can communicate with other computers by using common standards of data transmission.

Two types of port are fitted to computers – a parallel port (often used for printers), and one or more serial ports. Mice normally use one of these. It is normal to have two serial interfaces fitted to your computer, one for the mouse, the other for connection to a modem. Modems convert computer data into an audio form, compatible with the telephone network. The process is called MOdulating and DEModulating – hence the acronym.

Computer users make regular contact using modems. Data is passed between groups around the country and the world using this system. There are networks of computer users, often with specialist interests, such as satellite buffs, who are willing to share all manner of data via Bulletin Board Systems (BBS) set up specifically for this purpose.

Viruses

Viruses are programs designed to cause various types of problems by interfering with normal computer operations, often in an unpredictable manner. There are many types: Trojans, logic bombs, worms, boot virus, … the list goes on for thousands of types. Virus writing and dissemination is a criminal act. To prevent your computer being 'infected' with a virus, take common-sense precautions. Do not allow free access to your computer; don't use disks from unknown sources without careful checks; this is the most common method of virus transfer.

Purchase?

A knowledge of computer specifications enables one to understand the difference in capabilities between a BBC model B with 32 kb RAM, and the use of a modern PC with 4 Mb of RAM, a fast hard disk and SVGA monitor for displaying the highest resolution pictures.

When you have decided on the specification of your ideal system, you have a further choice. Buy locally or by mail order? You may pay extra, purchasing from a local supplier. His expenses become yours. He should, however, be able to help if you get really stuck.

Glancing at computer magazines shows some good deals are available, and credit cards can provide some protection. Don't be too shy about 'making an offer'! You may be pleasantly surprised.

Printers

Although not part of the computer, these are bought for production of printed material (hard-copy). In satellite work you may want to produce a 'screen dump' of pictures. Although this facility is usually provided by software, the quality obtained during graphics printing using a standard 9-pin dot matrix printer is not very good. A 24-pin printer gives better results, ink-jets and laser printers should provide the best.

The same printer may give different results depending on the printing software – that is the printer driver. This book includes pictures taken using screen dumps of the types mentioned.

The continuing fall in printer prices and increasing needs to produce good quality printed text mean that an ink-jet printer is to be recommended, assuming that a laser printer is outside your price range.

Lastly

I have not covered most peripheral devices – add-ons such as RTTY decoders, image scanners, and the multitude of other exciting things available for your computer.

Every user may eventually wish to upgrade their computer – perhaps to extend a particular aspect of their hobby. Some of the products that I have been fortunate to review on behalf of SWM readers show the increasing role that computers have to play in SW utility work.

Decoder facilities

Some satellite hobbyists may choose the framestore route because of its challenge, and the opportunities for later modification – assuming it is of the self-build type. For those choosing the computer decoding option, the choice becomes one of selection of a suitable unit from a number of suppliers. Such a choice can be made by reference to the following list of software options that can be considered to form an optimum list of facilities.

What you get

When you buy a weather satellite decoding unit, depending on the exact type, you can expect to receive the hardware as an expansion card to be slotted into your computer, with software to run it – supplied on a floppy disk. The PC GOES/WEFAX unit includes the interface on a small connector to plug into the serial port – installation could not be easier. Installation of expansion cards is normally straightforward, with software usually supplied as a self-installing batch file. For clarification, a 'batch-file' is a small program consisting of specific commands, each doing a separate job; this set of tasks forms a batch of jobs, e.g. copying files to a certain directory on the hard disk, expanding some files to their full size, and requesting user input.

To get you going without the need to provide live pictures to test, suppliers often include images previously obtained with their products. These enable you to experiment with facilities such as contrast expansion and adding artificial colour.

JVFAX

One notable exception to this type of image decoding is the JVFAX program suite. This comes as a nominally freeware program written by Eberhard Backeshoff, which, when used with a suitable hardware add-on, can decode both HF/Fax utility transmissions, and all the standard weather satellite formats. See Appendix 3 for details of this remarkable program.

Let us now look at the facilities that you might want your weather satellite decoding facility to have:

Optimum Facilities

- Polar orbiters
  - With sufficient (RAM) memory and a good quality monitor, your computer should provide excellent pictures in visible and infra-red, showing 64–256 grey levels. You may be able to store a whole pass, some 20 minutes in the case of METEOR class 3 satellites, and about 15 minutes of NOAA passes.
  - You should be able to set up the hardware to analyse the whole dynamic range of the audio signal from your receiver – normally one adjustment control is provided. After a pass you should be able to zoom into any area of the picture and resolve to the limit of the satellite's on-board sensors. When necessary, for example with METEOR pictures, you should be able to stretch contrast levels and reveal the detail. Land *is* visible in METEOR pictures if your software will let you see it.
  - The use of colour in infra-red (thermal) imagery is useful to reveal temperature gradations more clearly than grey scales. Using a sequence of colour changes – for example deep to light blue, and deep to light red, you can monitor (and measure) thermal variations in any image.
  - Synchronization of the picture may be done using the satellite's picture format by hardware or software. Each has its benefits and disadvantages. Software should be flexible enough to allow synchronization of noisy signals, and for the differing formats of OKEAN images. Suppliers should confirm their willingness to provide upgrades if new formats are used.
  - Pass scheduling may be possible if the software and receiver can cooperate!
  - Software should allow images to be saved in a recognized

format, as well as the program's native form. Well-known
formats include GIF, PCX and TIF. The reason for this is the
desirability of later image importing into other software – such
as DTP (desk-top publishing) – or into a proprietary image
processing program. If you want hard-copy (a print-out on
paper) you may need to export the raw image to a
word-processor. Most programs will allow importing of one or
more of the popular formats mentioned.

- METEOSAT
  Some of the above requirements also apply to METEOSAT
  imagery.
  - Software should permit the storage of complete picture
    information, each pixel containing one of 256 intensity levels,
    where appropriate. Certain image formats, for example water
    vapour images, do not contain this number of grey scale
    intensities.
  - There should be adequate 'zoom' facilities to examine detail.
  - You may want to program the software to store selected
    frames, identified by their scheduled transmission times.
  - Contrast adjustment is essential, and colour attribution is
    useful even when no image temperature calibration is available.
  - Channel changing (A1 and A2) may be selectable either
    manually or in software. A complete system should include this
    in software.
  - Facilities such as 'country outline removal' and 'noise
    reduction' are helpful where screen photography or image
    printing is required.
  - Animation:
    Transmissions by METEOSAT are done, as far as possible, by
    reference to a timetable described in Appendix 5. Certain
    picture formats for example Europe, the UK, and American
    re-transmissions, are always done at a specific time each hour,
    enabling repeated collection of image data of any specific area,
    e.g. the UK and Atlantic ocean. By sequencing these images, a
    realistic short 'film' (an animation) is produced showing regular
    changes of the weather patterns.
- Satellite tracking programs.

Whether the software that you buy includes a satellite predictions program, or
not, this is an essential part of your 'kit'. Such software rarely costs more than
£25 and you can obtain good software in the public domain or shareware
areas – see Appendix 4. You want to know when polar orbiting satellites are in
your vicinity, and you may also take an interest in other satellites.

Check software specifications

During recent years I have had the opportunity to test different decoding cards and software, so some brief product reviews are included in Appendix 4. Software will be updated from time to time so it is important to confirm that your final choice does offer those facilities, described previously, that you decide are important. Suppliers will normally cooperate in providing information, so an assortment of literature can be perused before purchase.

Image processing

Software used to decode weather satellite images should provide a facility to save the raw data; this option might use the programmer's own format (such as .NOA in the case of Timestep's polar orbiting software) or one of the standard image formats – .GIF, .PCX, .TIF amongst others. (Timestep also provide the PCX format option.) Whatever format is used, sooner or later you will want to enhance the image.

A good example of the need for such enhancement is shown by infra-red images from both METEOSAT and NOAA satellites. On most occasions during the day, thermal differences between cloud, land and sea will be enough to provide a reasonably contrasty image. Night-time infra-red pictures are often somewhat thin, exhibiting little temperature difference between land and sea.

Some decoding software allows examination of the image on a pixel/group of pixels basis. Poor contrast images tend to have the majority of pixels with similar values of intensity. Image-processing software lets us change the values of numbers of the pixels in order to stretch the contrast of the final picture. It is a scientifically valid process, though images so treated will no longer contain valid temperature information. This is not a problem because we usually leave the original image untouched, and work on a copy.

There are several forms of image processing. Images can be subjected to *histogram equalization*, in which the image is examined by the software, and those pixels having the most common values, have those values adjusted to fall within the middle range of intensities – other pixel intensities being adjusted accordingly. This process is very effective for bringing out otherwise hidden land detail in METEOR images. A *smoothing* filter checks for sudden changes in intensity and smoothes the group of pixels – effective in reducing the impact of noise in the image. *Edge enhance*ment improves the clarity of borders between areas of different intensities – best used on noise-free images! The sometimes annoying country outlines superimposed on METEOSAT images can be greatly reduced using a *median filter*.

9 Projects with weather satellites

The study of satellites, and reception of their data, involves many disciplines of science and technology. Electronic projects are evident, even if you choose to avoid receiver construction. Satellite orbits invite the study of mathematics; computer programs can be written to predict the pass times of satellites for your own location, and images received give opportunities to learn about ecology, meteorology and oceanography, for example.

Schools and satellite data

Astronomy in the curriculum

My own interest in satellite reception started from an interest in astronomy, and as explained, this is well catered for by the UoSATs. The long-awaited introduction of astronomy into the curriculum can benefit from the use of UoSAT data.

Geography and physics

Weather satellites provide an unending source of teaching material for geography, and METEOSAT must take the highest position (literally) as such a provider. It disseminates constant imagery of the continents on our side, and much of the other side of the globe, with virtually continuous, near-live pictures of Africa, Spain, France and the rest of Europe, and of course the UK. No atlas or film can compete with METEOSAT, yet how many schools have suitable receiving equipment, even assuming that staff are aware of the possibilities?

The European Space Agency is committed to maintaining METEOSAT for many years; it is a poorly equipped school that cannot show its pupils pictures from space that are so freely available.

METEOSAT provides pictures from GOES and GMS, and in 1991 METEOSAT-3 started transmitting from its temporary home at longitude 50° west. It was later moved to about 70° west, and replaced by the recently launched GOES-8 weather satellite, positioned at 75° west. Does anyone really doubt the enormous benefit to the study of geography provided by the reception of near real-time pictures of the Amazon, the Nile and Atlas mountains, tropical storms etc.?

Projects

Based on my own monitoring of METEOSAT and GOES pictures going back over several years, I have devised these projects for schools to consider. Teachers of these subjects may wish to modify the detail to suit their own course requirements.

Project 1: Iceberg formation in Bothnia

The months from May to October or November are almost always free of ice in the inland waterways near Sweden and Finland – the Gulf of Bothnia. The onset of cold weather causes ice to accumulate, usually in the upper Gulf, gradually increasing their surface area. It first has the appearance of persistent fog!

Winter often comes rapidly to this region but during 1989 and 1990 there was unusually mild weather and ice did not form to any large extent until near the end of winter, and did not persist.

An interesting, long-term project for geography lessons would be to monitor this region more closely, noting the start date of the first ice formation in the region and recording its progress, which should correlate with local temperatures.

The cheapest METEOSAT receiving system that can be bought may not be able to measure the temperature, in which case it will be necessary for someone to use teletext (from a suitable TV channel) on a daily basis, to find this information. Current decoding systems have the facility to calibrate METEOSAT frames and therefore enable this datum to be recorded.

Project 2: The advance and retreat of Greenland's ice

Like the South Pole, the North Pole is subject to the greatest extremes of exposure to the Sun. For six months of the year it is in continuous darkness, then for the remaining six months it experiences continuous sunshine. This inevitably causes the most dramatic effects to be seen along its coastline, where the ice is rarely stable – it will be either slowly melting or forming.

From Britain we can monitor areas near the North Pole, using the polar

Figure 9.1   *A METEOR picture showing ice around the Kola Peninsula, the White Sea and Upper Bothnia*

orbiters. There are always active NOAA weather satellites, and all are in high inclination orbits, passing close to the polar regions. During the summer months, with the north pole remaining in sunshine, the CIS METEOR satellites continue transmitting visible-light pictures as they pass over this region, and may transmit very good quality pictures.

If photographs or graphic images are produced at regular intervals during the year, it becomes apparent that there are strong seasonal changes taking place. As winter approaches, METEOR-3 series craft stop transmitting visible-light images because of the drop in light level, so monitoring must be done using the infra-red pictures from NOAA and any operating METEOR-3 infra-red transmitting satellites.

During February 1992 METEOR 2-19 was in a southbound orbit coming from over the North Pole. At mid-day there were very clear views of the ice formations all along the south-eastern coast of Greenland, giving a live view of the icebergs.

The type of ice formation also changes. As well as large icebergs persisting all winter, many small icebergs are seen, and with regular monitoring, one can follow the development and movements of these.

Because of the higher orbits of series three METEORS, we can follow even more of Greenland's ice sheets than when using the NOAAs. Using a vertical

Figure 9.2 *Ice near Greenland*

crossed dipole antenna, I can monitor the whole of Greenland on most westerly passes of METEOR 3-5 (when operating). If your antenna is steerable, or if you can point it northwards, you may improve your signal strength and so follow the satellite further to the north.

Monitoring these coasts over a period of a couple of years enables you to see how repetitive the melting process remains. The same coastal shapes return each year, though the stretches of ice sheet may vary, depending on the severity of the winter. Similarly the degree of melting depends on seasonal cloud cover and therefore the maximum temperatures reached during the summer period.

Project 3: The Italian Alps

One of the first areas noticed, when starting to regularly monitor satellite imagery, is that of the Italian Alps. This mountainous region is always covered with snow, and becomes instantly recognizable – a landmark.

Whatever season, we can expect to see the same formation, the only real change is in the amount of snow coverage on the mountain tops. METEOSAT's C03 scan includes Italy and Europe, so a weekly study of this scan can reveal the advance and retreat of snow cover during the seasons. Even in the hottest summers there is still a covering.

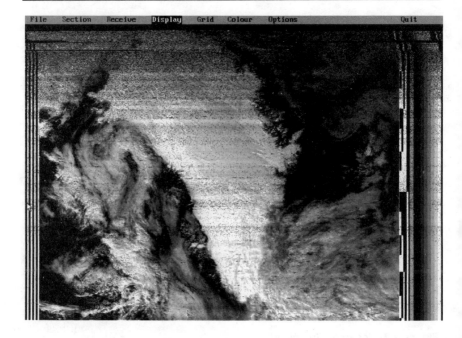

Figure 9.3   *METEOR 3.5 image of Greenland*

Project 4: Global cloud variations

Over a period of months, continuous monitoring of METEOSAT images reveals other events. When you see your first few pictures from METEOSAT, the novelty of seeing the world from a spacecraft some 36 000 km distance may amaze you – you cannot take in all the detail in one viewing. Study your first pictures in detail; you may notice several features not seen at first glance.

Look at a well-produced image of North Africa – the C2D format – see Fig. 9.5. It should contain some 64 grey levels, revealing several features. In the dark levels – the Atlantic and Mediterranean Sea – you should spot some islands. There may be some areas of cloud with similar dimensions. Zooming in (assuming you are using a computing system) should allow you to differentiate between the two. Land may show the greatest spread of grey levels; in Africa you can see the mountain ranges casting shadows in the morning or evening sun, and causing local cloud formations. The effect of vegetation becomes more pronounced in European regions – France, Spain and countries further east are generally darker.

Clouds may show a good range of grey levels, from the lower darker rain clouds to the higher, lighter, colder clouds, all in different formations.

Records of the average cloud cover over various regions can be kept. I see far fewer clouds over North African deserts than over the north Atlantic. African

Figure 9.4 *METEOSAT C03 format*

clouds seem to spring up over a few hours and disappear as quickly, often near the Atlas mountains. This forms a perfect geography study!

Project 5: Animating weather systems

The progress of depressions from west to east, across the Atlantic, then Europe, can be watched. Animated sequences are most effective. Using a METEOSAT computer system you can animate several frames, such as D2 or C02, and study short-term system movements.

After some hours spent studying cloud formation and distribution, you can notice certain repeating features; clouds hug the coast of Italy, running from north to south, but keeping within the land mass. Watch out for similar effects, particularly along the western coast of South America.

Project 6: The monitoring of thermal currents

Visible light pictures from NOAA and METEOSAT provide clear cloud, land and sea images; those from the METEORs show excellent cloud detail, though

Figure 9.5   *METEOSAT C2D format*

land can be revealed when the image is subjected to computer enhancement. A scan of the sea in visible light, clear of cloud, may appear uniformly dark (ignoring solar reflections).

As explained in Chapter 3, infra-red images are of a different nature. The on-board scanner is calibrated regularly and provides images where the amount of visible detail depends on temperature differences.

A day-time, infra-red image of the D2 format from METEOSAT includes North Africa and Europe, and shows a warm, and therefore dark desert, with cold (white) clouds. Compare this with METEOR (series 3 infra-red pictures), in which cold clouds appear black, and the warm Mediterranean Sea is white.

If we closely examine an infra-red image of the sea, we can identify a number of features. Temperature variations are seen throughout the image. The variations are real, but they must be interpreted carefully. Can you tell the difference between a sea temperature of 5°C and a cloud temperature of 5°C? They are indistinguishable! The image is showing temperature variations—whatever causes them; the skill lies in correctly identifying the source—cloud or sea. In practice, problems are not insurmountable, and on many occasions there will also be water vapour imagery available from METEOSAT and NOAA.

One can analyse images and monitor thermal variations in sea currents throughout the year. On one occasion I saw the dramatic exit of warm water from the Mediterranean Sea, no doubt a tidal effect. This was observed using a METEOSAT image; a later visible frame showed that the effect was a true sea-temperature variation – there were no clouds in that area to distort the measurements of the sea's temperature.

Project 7: Diurnal thermal changes

The warmth brought by the sun is demonstrated dramatically in infra-red images. To compare qualitative changes, we must analyse pictures from one satellite on a frequent basis. METEOSAT therefore becomes the prime contender for such a project.

One of the first thermal changes observed when starting weather animation, is that shown by D2 imagery. The warmth of the North African desert causes D2 images of that region to darken, and if a long sequence is recorded, the overnight cooling-down effect is quite marked. During the day-time there are large temperature changes from perhaps 10°C to over 50°C; night-time variations are much smaller.

This project concerns those changes that can be studied over various types of surface, vegetation, desert, sea etc., all of which exhibit different changes of temperature during the 24 hour period. Sandy deserts may show the largest diurnal variations, and the sea the smallest. Water has the highest specific heat – it holds its energy for longer. These changes may be greater during the summer than during the winter.

Project 8: Amazon – special studies

A study of the South American pictures taken by GOES-8 and re-transmitted by METEOSAT-5 allows us to examine the whole continent, and take a close-up view of the Amazonian forest – see Fig. 9.6. This area has been in the news for some years because of the rapid deforestation. It surprised me that so little coverage has been provided by the media of the views from METEOSAT.

If you monitor South American images regularly, you may notice features which appear permanent. My early recordings confirm later views showing large areas of either cloud or ground-based material hugging the western coast-line. Analysing infra-red alone is not convincing because, as previously pointed out, differences in thermal imagery can be cloud or ground cover; for unchallengeable evidence, we need visible imagery, and preferably water vapour pictures as well. You must always prepare your case well in scientific research. Try not to jump to conclusions on the basis of minimal evidence–that's the road to pseudo-science!

Using the numerous types of pictures available, separate projects for the study of the Amazon and other areas of interest can be identified. Changing the

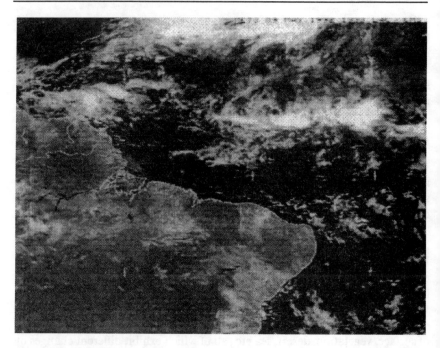

Figure 9.6   *GOES image of Amazon region*

environment, for example replacing vegetation with desert or concrete, usually changes its properties of thermal retention. From that observation, monitoring diurnal variations of the different areas of South America can help to indicate what the local surfaces are – and show any marked changes.

Project 9: Tropical hurricanes

These are perhaps the most dramatic weather systems and are easy to observe. Using METEOSAT's relayed images from the Americas, we can 'animate' American images.

During several years of weather satellite monitoring, I have seen hurricanes of differing intensities develop in the Florida and eastern Pacific regions; we can guarantee to record one sooner or later.

Figure 9.7 was taken on 22 September 1992 by METEOSAT-3, positioned over the eastern coast of America, showing a hurricane well into the Atlantic Ocean.

If you become known to the local media for satellite monitoring, you may get calls from reporters asking predictable questions such as 'Is this the biggest hurricane that you have ever seen?' and other leading questions. The press are apt to write articles with some embroidery; I would suggest trying to avoid too

Figure 9.7 *Hurricane season near North America*

much drama or things can really get out of hand. You may find it almost impossible to predict the movement and life of a tropical hurricane. The Met Office have incomparable resources and skills to do that sort of thing and we simply cannot compete. However, there are areas where the reverse is true – but that's for another chapter!

Project 10: Weather in areas of differing agriculture

This project is really worded backwards, since agriculture generally depends on the local weather, rather than the other way round. Studies of North America show the different prevailing weather in various regions. The prairies, deserts and lakes of the north all have seasonal variations. Vast areas of North America are used for growing wheat and herding cattle. Keeping records of local weather can reveal the well-known historical climates of the region.

Project 11: The oceans – seasonal changes

Seasonal changes can be monitored all over the visible part of the globe. The oceans have a different effect on the air mass above; monitoring cloud amounts over the north and south Atlantic during a period of some months reveals periods when cloud coverage is extensive, and other occasions when it is fairly sparse.

The Atlantic has deep underwater currents carrying warm water around the globe as part of a natural heat distribution system. On clear days one can see surface thermal currents as well.

There are well-known phenomena such as the El Nino effect in the Pacific Ocean, in which the normal movement of warm water is temporarily reversed, and this can cause severe natural disturbances to our weather. During 1990 the warm Gulf Stream moved from its normal position and we were hit by severe gales.

One project could include detailed monitoring of the oceans, recording of temperatures and cloud cover. For comparison purposes, historical data of this type is available from EUMETSAT in the form of CD-ROMs.

Summary

This chapter is not meant to be exhaustive – it is designed to scratch the surface and show how much information weather satellites can contribute to education. Teaching staff may see many other opportunities for incorporating this type of study into their lessons.

10 Sleuthing

If you have a satellite receiving station, with a scanner covering the 136.00 to 137.99 MHz band, and perhaps a second scanner covering wider bands of frequencies, you may try your hand at 'sleuthing' – the name given to the identification of satellite signals.

General satellite monitoring

Using a good tracking program, you can soon master the art/science of identifying various Commonwealth of Independent States (CIS) and other weather satellites during the times of their operations 'change-overs'. Earlier chapters mentioned that these satellites are not sun-synchronous – their orbital planes slowly change with respect to the Sun. By updating your tracking program monthly, you can monitor all these satellites. Just occasionally something amazing happens!

Several years ago, before I realized just how many old weather satellites were still in orbit (even if inactive), I started to monitor the band regularly. I came across a satellite (later identified as METEOR-30) transmitting on 137.02 MHz, as described in Chapter 4. This transmission was unexpected, so I referred it to more experienced people who had been monitoring for some years. Initially I was assured that it couldn't be METEOR-30 because 'the frequency is wrong'. Monitoring the band further, I started finding other unexpected transmissions – one regularly on 136.65 MHz – later identified as Transit 5B5.

On 18 August 1987 I received a call advising me of a new launch – METEOR 2-16. Returning to the receiver I found it locked on 137.30 MHz, and receiving a strange signal – so unlike a normal METEOR that further investigation was indicated. Others picking up the new satellite – METEOR 2-16 on 137.40 MHz – had missed this strange one on 137.30 MHz – just as I had missed METEOR 2-16. Following my alert, the strange signal was soon heard by others.

With access to a considerable number of Kepler elements of old Russian satellites, efforts were made to locate the source of the unusual transmission. It was identified as METEOR 2-3, an old weather satellite which had been in the same area of the sky when the new METEOR 2-16 was being commanded. It was most surprising to realize that the same command could switch on two different satellites! METEOR 2-3 was switched off after a few days' delay.

With considerable patience, you can hear a multitude of satellites in this band. Telemetry from METEOR 2-19 was partly blocked by a satellite called MAGION 3, operating on the same frequency (137.85 MHz). Fortunately the orbital period of MAGION 3 is such that only one pass per day was affected. On some occasions I have found three Russian satellites to be transmitting on the same frequency – all above my horizon at the same time!

With a suitable scanner and a good antenna – even a discone – you can hear a large number of satellites in nearby bands. I have discussed COSMOS navigation satellites operating in the 150 MHz band. These form a good test for receiving systems because there are so many operating. If you cannot hear them after trying for perhaps three hours, you must presume that your receiving equipment, probably your antenna, is insufficient.

A discone without a pre-amplifier is unlikely to receive the signals at sufficient level. My own discone is loft-mounted, and fitted with a pre-amplifier, prior to the 17 m cable run to the receiver. The top connector is an N-type, feeding UR67 cable, terminated in an N-type to BNC connector, which feeds the receiver. Although this is not perfect, I have been able to monitor satellites in most bands.

There are many other higher-frequency bands used by a variety of satellites, but these are more difficult to receive on basic equipment.

Caution

If you do make a 'sub' hobby of satellite sleuthing, you will soon realize that many apparent satellite signals are not genuine. I found several spurious signals associated with specific programs run on my computer! Different computers may produce different spurious frequencies, varying with the operation being performed by the computer. My receiver often locks on 137.62 MHz when I use my word processor! Similarly, the BBC model B (particularly its monitor) and many other computers can cause interference.

Minimizing interference

Interference can be minimized in a number of ways. Make sure that all of your equipment is properly earthed, and if possible, to the same earth point. It is possible to create an 'earth loop' in which a line of equipment has external cables linking signals together. This becomes susceptible to interference but this can be minimized by checking whether one or more earth return connections on the low signal voltage lines can be disconnected. It is essential to ensure that all mains earth connections are correct and secure. It is the earth braid on some signal cables that may be removed.

All antennae should be as far away from the house as possible. With a roof-mounted antenna and pre-amplifier (not a recommended addition) you will almost certainly pick up interference, often from signal break-through. Even without a pre-amplifier I sometimes hear taxis coming in 'on' a NOAA frequency. With a pre-amplifier, the situation would be almost hopeless. It is not that the taxis broadcast on a NOAA frequency, but because the pre-amplifier provides gain to a range of signals, and these reach the receiver's input. The input circuitry of any receiver will have filters incorporated, but for a scanner, these filters cannot be optimized and still work over a wide frequency range. There are, therefore, a large number of signals and harmonics reaching the receiver; even when tuning to a specific frequency, the receiver will let through some unwanted signals.

Because of this, great care must be taken when logging a signal that you believe to come from a satellite. The first step is to see whether the signal is associated with any domestic operations! Noting the start and finish times can be helpful and then attempts can be made to relate the observed frequency with known transmissions. This is where a comprehensive satellite frequency listing can be of use.

The unidentified signal may be heard again after a fixed period. Orbital periods vary enormously, but 100 minutes is common; remember you may have heard the last pass in a sequence. If you ever hear X3 (PROSPERO) on 137.56 MHz you will find that its orbit takes it over Britain several times per day from west to east. After these consecutive orbits there is a long gap of several hours. Identifying this satellite was not too difficult because of the frequency, previously used by earlier and later British satellites.

Known frequencies

With so many satellites transmitting on various frequencies, a book of this nature has to include a number of the most likely ones that you might come across, while monitoring with your scanner. Whether or not you actually hear them all depends on several factors:

- how much time you spend listening
- the effectiveness of your antennae, and the sensitivity of your receiver
- whether a particular satellite is actually transmitting near your location

In Chapter 2 several sets of frequencies were given for satellites in various bands. Table 10.1 gives a short list of some others not mentioned.

Table 10.1   *Some satellites and their frequencies*

Name	Frequency (MHz)
MOS 1, 1B	136.11
ORBCOMM FM	137.40, 137.68
INFORMATOR 1	145.815
DEBUT	136.89
DOVE 17	145.825
ALMAZ 1	179.875, 180.125

Conclusions

With hundreds of satellites in orbit, there are going to be many signals heard in these bands, often from long-discarded satellites. The summer months seem to produce more unexpected transmissions, often from old satellites that can occasionally operate in sunlight. Old, on-board batteries do not usually retain power after a few years have elapsed, so when a satellite is still transmitting telemetry, this may cease during the night-time part of its orbit.

The hobby of satellite sleuthing is an absorbing one and there is a sense of elation when a previously unrecognized satellite is identified. Remember that if your receiving system is not very efficient, you may not be able to monitor some bands very effectively, even if you can dial them on your scanner. You can test your equipment by listening to known transmissions.

I started this book wanting to provide an insight into the world of satellites. Through magazines and computer networks you can keep up to date with events, perhaps even contributing to the flow of news about satellite activities. Perhaps I shall see you on Starbase1 or even the Internet?

11 Late-breaking news from NOAA, GOES, GMS and GOMS

During the final stages of writing this book, a number of events occurred. New information about future plans for the polar orbiting weather satellites was released by NOAA; the newly launched GOES-9 and GOMS satellites provided their first images, and GMS-5 became operational. EUMETSAT commenced almost complete encryption of METEOSAT primary data images from September 1995; details are given here.

The following sections update the weather satellite scene as at autumn 1995.

- NOAA polar satellite update information
- Merging of NOAA and DMSP operations
- Geostationary update; GOES constellation update
- GMS and GOMS operations
- METEOSAT PDUS encryption
- Weather satellites and the Internet

NOAA Polar orbiters
NOAA 14 continues to transmit a.p.t. on 137.62 MHz. NOAA 12 transmits on 137.50 MHz. NOAAs 9, 10 and 11 recently ended routine operations.

NOAA-K: a brief overview of changes

The following information about NOAA-K onwards became available during May 1995.

In the Spring of 1996, a new series of operational environmental satellites will begin with the launch of NOAA-K. NOAA-K, L, and M will be the successors to the current NOAA operated, polar-orbiting satellites. They will carry a series of instruments which have been modified and improved from those now in orbit with the current operational satellites.

The advanced very high resolution radiometer (AVHRR/2) has been modified. The new instrument, AVHRR/3, adds a sixth channel in the near-IR, at 1.6 μm. This will be referred to as channel 3A and will operate during the daylight part of the orbit. Channel 3B corresponds to the previous channel 3 on the AVHRR/2 instrument, and will operate during the night portion of the orbit. The operational scheduling of the channel 3A/3B switching has not been precisely determined yet. A flag in word 22 of the telemetry will indicate which of the two channels is operating. Splitting channel 3 in this way maintains the HRPT data format which was designed to handle five AVHRR channels. Channels 3A and 3B are output at the same telemetry locations.

Automatic picture transmission (a.p.t.) users will receive the AVHRR/3 channel 3A, the same as channel 3B, with an ID wedge equivalent to grey scale wedge 3.

The microwave sounding unit (MSU) and stratospheric sounding unit (SSU) instruments have been deleted. NOAA-K will fly with advanced microwave sounding units AMSU-A1, AMSU-A2 and AMSU-B. The AMSU-A is a 15-channel microwave radiometer in two separate units. For AMSU-A1, -A2, the word 0001H will be used as fill data most of the time in the telemetry stream. The new AMSU data is expected to provide improved temperature and humidity soundings. Additionally, window channels 1, 2 and 15 will provide information on precipitation, sea ice and snow cover. The AMSU-B is a five-channel microwave radiometer; three of the channels are centred on the 183.31 GHz water vapour line. The other two channels are at 89 GHz and 150 GHz.

The solar backscatter ultraviolet radiometer (SBUV) is carried in satellites with an afternoon orbit. NOAA-K will likely be launched into an afternoon orbit and will carry the SBUV/2, which has only minor changes from similar instruments carried on previous spacecraft.

While NOAA-K will be tested in an afternoon orbit configuration, it will be capable of being launched as a morning or afternoon spacecraft to meet operational needs.

The new high-resolution infrared radiation sounder (HIRS/3) will have the calibration sequence changed. On HIRS/2, the calibration mode required the use of three calibration targets (space view, cold target, and warm target). On HIRS/3, the cold target will not be routinely used in the calibration sequence, resulting in one additional scan line of Earth data (38 Earth scans per 256 second cycle).

The Argos data collection system (DCS) aboard the NOAA polar orbiting satellites will be improved for NOAA-K. The DCS/2 will have an increased data transmission rate (from 1200 to 2560 bits per second) and the onboard data recovery units (DRUs) will be increased from four to eight.

An improved space environment monitor (SEM-2) has added in-flight calibration capabilities and improved particle detection. The total energy detector (TED) will measure to a lower energy (0.05 keV versus 0.3 keV on NOAA-J). The medium energy proton/electron detector (MEPED) has a fourth, omnidirectional, proton sensor for greater than 140 MeV. More data will be included in the telemetry stream.

The search and rescue processor (SARP) has added capabilities for the handling of distress messages, as well. The number of DRUs has been increased from two to three.

This information is provided to users of NOAA polar orbiting satellite data as an early indication of what changes they may expect with the launch of NOAA-K. However, it is preliminary in nature, and subject to revision prior to NOAA-K becoming operational in 1996.

The merging of NOAA and DMSP satellite programmes

On 1 June 1995, NOAA issued a statement (95-36) formalizing the establishment of the new civil-military satellite programme. The text is shown here:

NOAA 95-36
CONTACT: Patricia Viets, NOAA   FOR IMMEDIATE RELEASE
(301) 457-5005   6/1/95
Douglas Isbell, NASA
(202) 358-1753
LtCol Dave Simms, DOD
(703) 697-5131
AGENCIES ESTABLISH NEW CIVIL-MILITARY SATELLITE PROGRAM

The Clinton Administration has taken a major step toward combining the country's military and civilian weather satellite programs into a single system – a move that is expected to save taxpayers up to $300 million through 1999 with additional savings through the life of the program.

Secretary of Commerce Ronald H. Brown, Secretary of Defense William J. Perry, and NASA Administrator Daniel S. Goldin signed a formal agreement on May 26, establishing the agencies' roles and responsibilities in support of the new system, and implementing a Presidential Decision Directive that was signed last year.

'Combining these programs was a key recommendation of Vice President Gore's National Performance Review,' said Under Secretary of Commerce for Oceans and Atmosphere D. James Baker. 'The new program will result in a major reduction of acquisition, operational and facilities costs.'

Currently four U.S. polar-orbiting satellites are used to collect operational meteorological, oceanographic, climatic, and space environment data. Two satellites are provided and operated by the Department of Commerce's National Oceanic and Atmospheric Administration (NOAA), and two by the Department of Defense's Defense Meteorological Satellite Program. The new combined program will consist of three satellites. The first satellite under the new system, called the National Polar-orbiting Operational Environmental Satellite System (NPOESS), is expected to be launched in 2006.

To acquire and operate the NPOESS, the Department of Defense, NASA, and NOAA have established an Integrated Program Office. James T. Mannen, a retired Air

Force colonel with extensive experience in space programs, was named director of the office on May 30.

The signing of the agreement by the three agencies represents a tangible and significant step forward in interagency cooperation – merging operational military and civilian systems, while still satisfying each agency's critical mission requirements and doing so at reduced cost to American taxpayers.

My thanks to NOAA for providing this information.

Geostationary update

GOES constellation update

Operations using METEOSAT-3 formally ended on 31 May 1995 and GOES-8 continued routine transmissions from its easterly position. METEOSAT-3 was being drifted east to a backup position. GOES-8 can be received from more westerly counties, so its schedule is included here. The successful launch of GOES-J (9) returns the USA to its nominal operational status.

GOES-8 schedule

The new GOES-8 weather satellite schedule is published here, and is available from the Internet, on the NOAA.SIS web server. It includes both infra-red and visible-light transmissions from sections of the globe, as seen from above the east coast. Additionally, GOES transmits selections from METEOSAT-5, in a similar manner to the GOES pictures re-transmitted by METEOSAT-5 in the LY, LR and LZ slots. Also included are ice charts and FAX weather forecast charts. Images from NOAA-14, taken during its crossing of the poles, are included in the schedule, in both visible and infra-red formats in Mercator projection. Other sequences of forecasts are included, as well as the daily operations schedule, and, of course, TBUS data.

FIRST FULL DISK GOES-9 VISIBLE 12 JUNE 1995 17:45 Z (SSEC:UW-MADISON)

Figure 11.1 *First full-disk image from GOES-9, transmitted from its temporary location at 90° west. Image courtesy NOAA*

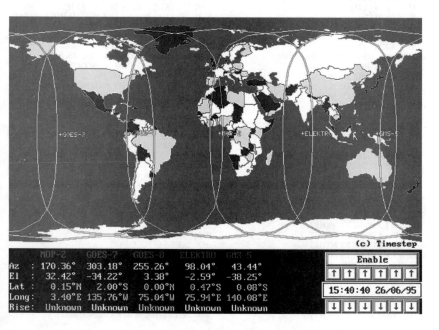

Figure 11.2 *MOP, GOES, ELEKTRO and GMS geostationary weather satellites—positions at mid-June 1995*

EAST WEFAX (GOES-8) SCHEDULE – 08/01/95

XMIT PRODUCT

0002	0450 MET-5 0430Z E8 MOIST
0006 MET-5 0000Z D3 IR	0454 MET-5 0430Z E9 MOIST
0010 MET-5 0000Z D4 IR	0458 GOES-8 0345Z US IR
0014 MET-5 0000Z D5 IR	0514 NOAA-12 POLAR VIS NH 080W-170W W034
0018 MET-5 0000Z D6 IR	0518 NOAA-12 POLAR VIS SH 080W-170W W035
0022 MET-5 0000Z D7 IR	0522 NOAA-12 POLAR NIR NH 080E-010E W036
0026 MET-5 0000Z D8 IR	0526 NOAA-12 POLAR NIR SH 080E-010E W037
0030 MET-5 0030Z D2 IR	0530 NOAA-12 POLAR VIS MER 100W-170W W038
0034 MET-5 0030Z D9 IR	0534 NOAA-12 POLAR NIR MER 080E-010E W039
0038 MET-5 0030Z D1 IR	0538 NOAA-12 POLAR DIR NH 080W-170W W040
0042 MET-5 0030Z D3 IR	0542 NOAA-12 POLAR DIR SH 080W-170W W041
0046 GOES-8 2345Z US IR	0546 NOAA-12 POLAR NIR MER 140E-070E W032
0050 W500 48 HR 250MB HT/TEMP/WND	0550 NOAA-12 POLAR DIR MER 040W-110W W033
0100 W501 72 HR SLP/1000-500TK	0558 GOES-8 0445Z US IR
0105 W502 72 HR 500MB HT/TEM/WND	0606 MET-5 0600Z C03 VS
0110 GOES-8 2345Z NE IR	0610 MET-5 0600Z D1 IR
0114 GOES-8 2345Z SE IR	0614 MET-5 0600Z D3 IR
0118 GOES-8 2345Z NW IR	0618 MET-5 0600Z D4 IR
0122 GOES-8 2345Z SW IR	0622 MET-5 0600Z D5 IR
0126	0626 MET-5 0600Z D6 IR
0130 GOES-8 2345Z US WV	0630 MET-5 0630Z D2 IR
0134 GOES-8 2345Z NE WV	0634 MET-5 0630Z C02 VS
0138 GOES-8 2345Z SE WV	0638 MET-5 0630Z C03 VS
0142 GOES-8 2345Z NW WV	0642 MET-5 0630Z C3D VS
0146 GOES-8 2345Z SW WV	0646 MET-5 0630Z C2D VS
0150	0650 MET-5 0630Z D3 IR
0154 GOES-8 0045Z US IR	0654 MET-5 0630Z D1 IR
0158	0658 GOES-8 0545Z US IR
0202	0702 GOES-8 0545Z NE IR
0210 NOAA-12 POLAR VIS NH 010E-080W W026	0706 GOES-8 0545Z SE IR
0214 NOAA-12 POLAR NIR SH 010E-080W W027	0710 GOES-8 0545Z NW IR
0218 NOAA-12 POLAR DIR NH 010E-080W W028	0714 GOES-8 0545Z SW IR
0222 NOAA-12 POLAR DIR SH 010E-080W W029	0718
0226 NOAA-12 POLAR DIR MER 010E-060W W030	0722
0230 NOAA-12 POLAR VIS MER 040W-110W W031	0726 NOAA-12 POLAR DIR MER 100W-170W W042
0235 GOES-8 0145Z US IR	0730 NOAA-12 POLAR DIR NORTH POLE W043
0240 ICE CHART	0734 NOAA-12 POLAR DIR SOUTH POLE W044
0245 ICE CHART	0738 GOES-8 0645Z US IR
0250 ICE CHART	0750 W006 24 HR PRECIP ACCUM VT00Z
0255 ICE CHART	0755 W007 24 HR SFC/1000-500THK
0300 WO64 SIG WX PROG FL250-600	0800 W008 24 HR 500MB HT/WD/TMP
0305 SIG WX PROG FL250-600	0805 W009 24 HR 250MB HT/WD/TMP
0310 MET-5 0300Z D4 IR	0810 W014 24 HR 250 MB PROG.
0314 MET-5 0300Z D5 IR	0815 W015 24 HR 250 HT ISOTACHS
0318 MET-5 0300Z D6 IR	0820 W016 24 HR TROP PRESS/VWS
0322 MET-5 0300Z D7 IR	0825 W017 24 HR TROP PRESS/VWS
0326 MET-5 0300Z D8 IR	0830 W018 24 HR 300 STM/ISOTACHS
0330 MET-5 0330Z D2 IR	0835 W019 24 HR 300 STM ISOTACHS
0334 MET-5 0330Z D9 IR	0840 W020 24 HR 200 STM ISOTACHS
0338 MET-5 0330Z D1 IR	0845 W021 24 HR 200 STM ISOTACHS
0342 GOES-8 0245Z US IR	0850 W022 24 HR 500 STM ISOTACHS
0346 GOES-8 0245Z NE IR	0855 W023 24 HR 500 STM ISOTACHS
0350 GOES-8 0245Z SE IR	0900 W010 24 HR 850 STM ISOTACHS
0354 GOES-8 0245Z NW IR	0905 W011 24 HR 850 STM ISOTACHS
0358 GOES-8 0245Z SW IR	0910 MET-5 0900Z D1 IR
0402	0914 MET-5 0900Z D3 IR
0406	0918 MET-5 0900Z D4 IR
0410 MET-5 0400Z E1 MOIST	0922 MET-5 0900Z D5 IR
0414 MET-5 0400Z E2 MOIST	0926 MET-5 0900Z D6 IR
0418 MET-5 0400Z E3 MOIST	0930 MET-5 0930Z D2 IR
0422 MET-5 0400Z E4 MOIST	0934 MET-5 0930Z C02 VS
0426 MET-5 0400Z E5 MOIST	0938 GOES-8 0745Z US IR
0430 MET-5 0430Z D2 IR	0945 W012 24 HR 700 STM/ISOTACHS
0434 MET-5 0430Z D1 IR	0950 W013 24 HR 700 STM ISOTACHS
0438 MET-5 0430Z D3 IR	0955 W100 00Z MSL PRES/1000-500TK
0442 MET-5 0430Z E6 MOIST	1000 W101 00Z MSL PRES/1000-500TK
0446 MET-5 0430Z E7 MOIST	1005 W102 24HR MSL PRES/1000-500TK

1010 W103 24HR MSL PRES/1000-500TK	1518 MET-5 1500Z D4 IR
1015 W104 48HR MSL PRES/1000-500TK	1522 MET-5 1500Z D5 IR
1020 W105 48HR MSL PRES/1000-500TK	1526 MET-5 1500Z D6 IR
1026 GOES-8 0845Z US IR	1530 MET-5 1530Z D2 IR
1030 GOES-8 0845Z NE IR	1534 MET-5 1530Z C02 VS
1034 GOES-8 0845Z SE IR	1538 GOES-8 1445Z US IR
1038 GOES-8 0845Z NW IR	1542 GOES-8 1445Z NE IR
1042 GOES-8 0845Z SW IR	1546 GOES-8 1445Z SE IR
1046 GOES-8 0945Z US IR	1550 GOES-8 1445Z NW IR
1050	1554 GOES-8 1445Z SW IR
1054 SCHEDULE FILE PART-1	1558
1058 SCHEDULE FILE PART-2	1602 GOES-8 1445Z US VS
1102 WEFAX MESSAGE FILE	1606 GOES-8 1445Z NE VS
1106	1610 GOES-8 1445Z SE VS
1110 W503 48 HR SLP/1000-500TK	1614 GOES-8 1445Z NW VS
1115 W504 48 HR 500MB HT/TMP/WND	1618 GOES-8 1445Z SW VS
1122 NOAA-12 POLAR VIS NH 170W-100E W001	1622
1126 NOAA-12 POLAR VIS SH 170W-100E W002	1626 GOES-8 1545Z US IR
1130 NOAA-12 POLAR NIR NH 010E-080W W003	1630 GOES-8 1545Z US VS
1134 NOAA-12 POLAR NIR SH 010E-080W W004	1638
1138 NOAA-12 POLAR VIS MER 170W-120E W005	1642
1142 NOAA-12 POLAR NIR MER 020E-050W W006	1706 NOAA-12 POLAR VIS MER 130E-060E W010
1146 NOAA-12 POLAR DIR NH 170W-100E W007	1710 NOAA-12 POLAR NIR MER 040W-110W W011
1150 NOAA-12 POLAR DIR SH 170W-100E W008	1714 NOAA-12 POLAR DIR MER 130E-060E W012
1154 NOAA-12 POLAR DIR MER 170W-120E W009	1718 NOAA-12 POLAR VIS NH 100E-010E W013
1158 GOES-8 1045Z US IR	1722 NOAA-12 POLAR VIS SH 100E-010E W014
1206 MET-5 1200Z C03 VS	1726 NOAA-12 POLAR NIR NH 080W-170W W015
1210 MET-5 1200Z D1 IR	1730 NOAA-12 POLAR NIR SH 080W-170W W016
1214 MET-5 1200Z D3 IR	1734 NOAA-12 POLAR VIS MER 070E-000E W017
1218 MET-5 1200Z D4 IR	1738 NOAA-12 POLAR NIR MER 100W-170W W018
1222 MET-5 1200Z D5 IR	1742 GOES-8 1645Z US IR
1226 MET-5 1200Z D6 IR	1746 GOES-8 1645Z US VS
1230 MET-5 1230Z D2 IR	1758 MET-5 D2 1800Z D2 IR
1234 MET-5 1230Z C02 VS	1802 MET-5 D1 1800Z D1 IR
1238 MET-5 1230Z C03 VS	1806 MET-5 D3 1800Z D3 IR
1242 MET-5 1230Z C3D VS	1810 MET-5 D4 1800Z D4 IR
1246 MET-5 1230Z C2D VS	1814 MET-5 D5 1800Z D5 IR
1250 MET-5 1230Z C1D VS	1818 MET-5 D6 1800Z D6 IR
1254 MET-5 1230Z D1 IR	1822 MET-5 D7 1800Z D7 IR
1258 GOES-8 1145Z US IR	1826 MET-5 D8 1800Z D8 IR
1302 GOES-8 1145Z NE IR	1830 MET-5 D2 1830Z D2 IR
1306 GOES-8 1145Z SE IR	1834 MET-5 D9 1830Z D9 IR
1310 GOES-8 1145Z NW IR	1838 GOES-8 1745Z US IR
1314 GOES-8 1145Z SW IR	1842 GOES-8 1745Z NE IR
1318	1846 GOES-8 1745Z SE IR
1322 GOES-8 1145Z US WV	1850 GOES-8 1745Z NW IR
1326 GOES-8 1145Z NE WV	1854 GOES-8 1745Z SW IR
1330 GOES-8 1145Z SE WV	1858
1334 GOES-8 1145Z NW WV	1902 GOES-8 1745Z US VS
1338 GOES-8 1145Z SW WV	1906 GOES-8 1745Z NE VS
1342 GOES-8 1145Z US VS	1910 GOES-8 1745Z SE VS
1346	1914 GOES-8 1745Z NW VS
1350 GOES-8 1245Z US IR	1918 GOES-8 1745Z SW VS
1354 GOES-8 1245Z US VS	1922
1358	1926 GOES-8 1845Z US IR
1402	1930 GOES-8 1845Z US VS
1410 TBUS NOAA-9	1938
1415 TBUS NOAA-10	1942
1420 TBUS NOAA-11	1950 NOAA-12 POLAR DIR NH P 100E-010E W019
1425 TBUS NOAA-12	1954 NOAA-12 POLAR DIR SH P 100E-010E W020
1430 W505 48 HR 250MB HT/TMP/WND	1958 NOAA-12 POLAR DIR MER 070E-000E W021
1435 GOES-8 1345Z US IR	2005 W047 24 HR SFC/1000/500THK
1440 GOES-8 1345Z US VS	2010 W048 24 HR 500MB HT/WD/TMP
1445 W506 72 HR SLP/1000-500TK	2015 W049 24 HR 250MB HT/WD/TMP
1450 W507 72 HR 500MB HT/TMP/WND	2020 W054 250 HT ISOTACHS
1455 W508 72 HR 250MB HT/TMP/WND	2025 W055 250 HT ISOTACHS
1500 W066 SIG WX PROG FL250-600	2030 W056 24HR TROP PRES/VWS
1505 SIG WX PROG FL250-600	2035 W057 24HR TROP PRES/VWS
1510 MET-5 1500Z D1 IR	2040 W058 24HR 300 STM/ISOTACHS
1514 MET-5 1500Z D3 IR	2045 W059 24HR 300 STM/ISOTACHS

2050	W060 24HR 200 STM/ISOTACHS	2230	GOES-8 2045Z US IR
2055	W061 24HR 200 STM/ISOTACHS	2234	GOES-8 2045Z NE IR
2100	WO62 24HR 500 STM/ISOTACHS	2238	GOES-8 2045Z SE IR
2105	W063 24HR 500 STM/ISOTACHS	2242	GOES-8 2045Z NW IR
2110	GOES-8 1945Z US IR	2246	GOES-8 2045Z SW IR
2114	GOES-8 1945Z US VS	2250	
2118	MET-5 2100Z D6 IR	2254	GOES-8 2045Z US VS
2122	MET-5 2100Z D7 IR	2258	GOES-8 2045Z NE VS
2126	MET-5 2100Z D8 IR	2302	GOES-8 2045Z SE VS
2130	MET-5 2130Z D2 IR	2306	GOES-8 2045Z NW VS
2135	W050 24HR 850 HT ISOTACHS	2310	GOES-8 2045Z SW VS
2140	W051 24HR 850 HT ISOTACHS	2314	
2145	W052 24HR 700 HT ISOTACHS	2320	W509 48 HR SLP/1000-500TK
2150	W053 24HR 700 HT ISOTACHS	2325	W510 48 HR 500MB HT/TMP/WND
2155	W150 00Z MSL PRES/1000-500TK	2330	NOAA-12 POLAR VIS NH 010E-080W W022
2200	W151 00Z MSL PRES/1000-500TK	2334	NOAA-12 POLAR VIS SH 010E-080W W023
2205	W152 24HR MSL PRES/1000-500TK	2338	NOAA-12 POLAR NIR NH 170W-100E W024
2210	W153 24HR MSL PRES/1000-500TK	2342	NOAA-12 POLAR NIR SH 170W-100E W025
2215	W154 48HR MSL PRES/1000-500TK	2346	GOES-8 2145Z US IR
2220	W155 48HR MSL PRES/1000-500TK	2350	GOES-8 2145Z US VS
2225	W156 24HR PRECIP ACCUM VT 12Z	2354	GOES-8 2245Z US IR
		2358	.

GOES-9 becomes operational

Following the successful launch of GOES-J on 31 May 1995, it completed the last major orbital manoeuvre to achieve semi-geosynchronous orbit. The solar arrays were partially deployed a couple of hours after launch, and the satellite was formally renamed GOES-9 the same day. Hours later, the solar array was further deployed to half-normal status, then the imager and sounder were switched on. The GOES spacecraft was allowed to outgas before the detector cooler covers were opened.

Checkout for GOES-9 was carried out at longitude 90° west. Post-launch tests were completed during summer. GOES-9 is to become the new GOES-West spacecraft.

The following text includes extracts from the GOES Mission Summary issued in June, excluding some paragraphs which have already been detailed.

GOES mission overview

Over the past 30 years, environmental service agencies have stated a need for continuous, dependable, timely and high-quality observations of the Earth and its environment. The new generation Geostationary Operational Environmental Satellites (GOES I–M) provide half-hourly observations to fill the need. The instruments on board the satellites measure Earth-emitted and reflected radiation from which atmospheric temperature, winds, moisture, and cloud cover can be derived.

The GOES I–M series of satellites is owned and operated by the National Oceanic and Atmospheric Administration (NOAA). The National Aeronautics and Space Administration (NASA) manages the design, development, and launch of the spacecraft. Once the satellite is launched and checked out, NOAA assumes responsibility for the command and control, data receipt, and product generation and distribution.

Each satellite in the series carries two major instruments: an Imager and a Sounder. These instruments resolve visible and infrared data, as well as temperature and moisture profiles of the atmosphere. They continuously transmit these data to ground terminals where the data are processed for rebroadcast to primary weather services both in the United States and around the world, including the global research community.

The GOES I–M mission is scheduled to run from the mid-1990s into the first decade of the 21st century. Each element of the mission has been designed to meet all in-orbit performance requirements for at least five years.

The GOES I–M system performs the following basic functions:

- Acquisition, processing, and dissemination of imaging and sounding data
- Acquisition and dissemination of Space Environment Monitor (SEM) data
- Reception and relay of data from ground-based data collection platforms (DCPs) that are situated in carefully selected urban and remote areas to the NOAA Command and Data Acquisition (CDA) station
- Continuous relay of Weather Facsimile (WEFAX) and other data to users, independent of all other functions
- Relay of distress signals from people, aircraft, or marine vessels to search and rescue ground stations of the Search and Rescue Satellite Aided Tracking (SARSAT) system

The GOES I–M system serves a region covering the central and eastern Pacific Ocean; North, Central, and South America; and the central and western Atlantic Ocean. Pacific coverage includes Hawaii and the Gulf of Alaska. A common ground station, the CDA station located at Wallops, Virginia, supports the interface to both satellites. The NOAA Satellite Operations Control Center (SOCC), in Suitland, Maryland, provides space-craft scheduling, health and safety monitoring, and engineering analyses.

Delivery of products involves ground processing of the raw instrument data for radiometric calibration and Earth location information, and retransmission to the satellite for relay to the data user community. The processed data is received at the control centre and disseminated to the National Weather Service's (NWS) National Meteorological Center, Camp Springs, Maryland, and NWS forecast offices, including the National Hurricane Center, Miami, Florida, and the National Severe Storms Forecast Center, Kansas City, Missouri. Processed data are also received by Department of Defense installations, universities, and numerous private commercial users.

The GOES ground system

The ground system supplies two major functional areas: (1) spacecraft and instrument health and safety monitoring, commanding and operations analysis and (2) Imager and Sounder instrument data processing. For each operational GOES spacecraft, the ground system ingests the raw instrument sensor data stream and generates a processed data stream that is transmitted back to the GOES for rebroadcast to the primary weather system users. For each operational spacecraft, the ground system also provides synchronous orbit and attitude determination to support the GOES system image navigation and registration function. Additionally, for each operational or standby GOES, the ground system provides orbit predictions, manoeuvre planning and commanding, and telemetry processing and analysis to support daily and periodic satellite operations. The ground system subsystems are located at the NOAA CDA station at Wallops, Virginia, and the SOCC at Suitland, Maryland.

The Weather Facsimile Service facilitates the retransmission of images and meteorological analysis from the Wallops CDA ground station to the user community. Data originates from the NWS and NOAA image processing facilities.

The Data Collection System (DCS) provides for near real-time acquisition and relay of environmental data for centralized archiving and distribution. The data is used to provide warnings and forecasts of environmental events such as tsunamis, tropical cyclones, and floods. It also is used to map river states, soil conditions and snow depth.

The GOES I–M satellites provide continued support for the communications paths between Earth-situated DCPs in the western hemisphere on 200 domestic and 100 international channels. The spacecraft can be programmed to support DCS operating frequencies of either the Japanese or the European geostationary satellites. As an adjunct to the DCS capability, GOES will continue to relay a National Institute of Standards and Technology time code for users throughout the hemisphere.

For further information, contact the Public Affairs Offices at either NASA (Maryland, USA) or NOAA (Washington, USA):

Public Affairs Office
NASA/Goddard Space Flight Center
Greenbelt, Maryland 20771
(301)286-6255

Public Affairs Office
NOAA
National Environmental Satellite, Data, and
   Information Service (NESDIS)
FB-4, Room 0124
Washington, DC 20233
(301)763-2560

GMS update

The Japanese geostationary weather satellite GMS-5 went into operation at 0532UTC on 13 June 1995, when it was located at 141.9° east, on its way to 140.0° east, scheduled for 0500UTC on 15 June. GMS-4 is now non-operational and drifting towards the back-up position at 120° east. Although not directly receivable from UK shores, this information is provided for completeness.

GMS-5 has one new infrared micron band added (6.5–7.0) (water vapour). The infrared thermal channel has been split in two (10.5–11.5, 11.5–12.5).

Its WEFAX (SDUS) specifications include the following:

- frequency 1691.0 MHz
- modulation: AM
- bandwidth 260 kHz
- output signal 2.4 kHz

The low-resolution imagery (WEFAX) is at 1691.0 MHz, and is a relatively standard WEFAX signal (AM modulation of a 2.4 KHz subcarrier, which FM modulates the r.f.). It has one unusual characteristic – the r.f. bandwidth is 260 kHz.

My grateful thanks to Kjell Magnussen for some of this information.

GOMS

The recently launched CIS geostationary WXSAT GOMS, now called ELEKTRO, is positioned over India at longitude 76° east, but is not transmitting regular telemetry. My thanks to Mike Kenny of Satellite Engineering, Bureau of Meteorology, Melbourne, Australia for providing this information.

METEOSAT data encryption

The encryption of METEOSAT Primary Data was implemented progressively during 1995. To decode the data stream, users require a METEOSAT Key Unit (MKU) which costs 700 ECU (approximately £500). An additional requirement is an interface unit for the computer, and this will depend on the decoding system provided by the manufacturer. Charges for such units are bound to vary, but can be expected to cost from £100.

For education and private users, subject to individual decisions being made by the relevant authority, an annual fee is unlikely to be charged. Under these circumstances, if you buy the necessary hardware and have approval from the Meteorological Office/EUMETSAT, you should be permitted to receive all primary data images from METEOSAT.

Without any additional equipment, you should be able to receive the unencrypted images which are to be transmitted every six hours.

Weather satellites and the Internet

The interconnecting computer network known as the Internet is a most valuable resource in many walks of life. For the satellite enthusiast it can be a very useful facility. However, it is perfectly true that you can obtain all the information required in order to pursue your own specialist interest, without being attached to the Internet. Kepler elements, tracking software and regular information are all obtainable through reasonably conventional sources, such as amateur groups – see Appendix 1.

If you have access to the Internet, the following sites may contain information of interest in your field. Please remember that site addresses occasionally change; all these have been used regularly by me during recent weeks.

World-Wide-Web sites

- NASA has many web pages of which several are linked together, and to other sites, enabling you to cross the continents with ease:

  http://www.gsfc.nasa.gov   their home page
  http://www.nasa.gov
  http://www.stsci.edu/EPA/Recent.html   Hubble Space
                                         Telescope results
  http://www.ksc.nasa.gov/shuttle/countdown
- NOAA also operates various pages of which the following provide current information on the weather satellite status.

  http://www.noaa.gov   Home page
  http://psbsgi1.nesdis.noaa.gov:8080
  http://arc.iki.rssi.ru   Russian Space Research Institute
- CERN – where it all started

  http://info.cern.ch   start here to go anywhere!

File transfer protocol (ftp)

The following sites can be accessed using ftp and usually have a large number of satellite related data and images.

nic.funet.fi/pub/ham/satellite/orbits   recent two-line elements from Finland

power.ci.uv.es/pub/satellites   another source of elements
ftp.ngdc.noaa.gov   NOAA information
smis.iki.rssi.ru   Russian data
ftp://spacelink.msfc.nasa.gov   large amount of teaching related material

From time to time, I provide new site addresses in my column 'Info in Orbit' in *Short Wave Magazine.*

Appendix 1 Useful addresses and sources of information

Manufacturers and retailers

The following companies (listed alphabetically) have provided information on products related to the weather satellite market. In a few cases the equipment is in the form of kits. You are recommended to study the appropriate sections in this book before contemplating the purchase of any product which may require some expertise in its construction or assembly.

This information has been collected from the companies listed, sometimes without product testing. It is essential that, before purchasing a product, you check with the supplier that the product will do what you expect of it. Product specifications and prices can change over a period of time, so prices must be taken as providing a guide.

- Centre for Satellite Engineering Research, University of Surrey, Guildford, Surrey, GU2 5XH. Tel (01483) 509131/509143.
- Cirkit Distribution Ltd, Park Lane, Broxbourne, Herts, EN10 7NQ. Tel (01992) 444111.
  Retails kits and some completed modules.
- Dartcom, Powdermills, Postbridge, Yelverton, Devon, PL20 6SP. Tel (01822) 88253.
  Retails products supplied for the professional market.
- Garex Electronics, Station Yard, South Brent, South Devon, TQ10 9AL. Tel (01364) 72770.
- Maplin Electronics, PO Box 3, Rayleigh, Essex, SS6 2BR. Tel (01702) 554161.
  Kits and complete units available.
- Martelec Communication Systems, The Acorns, Wyck Lane, East Worldham, Alton, Hants, GU34 3AW. Tel (01420) 82752.
  Supplies AMIGASAT 3.2 for Commodore Amiga computer; decodes all weather satellites, digital software processing. JVFAX interface around

£80; MSC30 a Meteosat downconverter; various wxsat products.
- Spacetech, 21 West Wools, Portland, Dorset, DT5 2EA. Tel (01305) 822753.
  Retails Atari 520ST to Mega4 ST computers: Weather satellite decoder; comprehensive software, METEOSAT animation etc.; specializing in schools' supplies.
- Timestep Weather Satellite Systems, PO Box 2001, Newmarket, CB8 8QA. Tel (01440) 820040.
  PROSAT II: hardware/software for PC computers (80286 and above) for the collection of all types of a.p.t. satellite transmissions; METEOSAT, NOAA, OKEAN, METEOR. See review section: Price £99 inc.
  Also PDUS systems for primary METEOSAT data. Enquire for current prices. HRPT systems for high resolution picture telemetry; Enquire for current prices.

Useful literature

The following list provides material for further reading on satellites and space matters.

- *Practical Wireless*, Arrowsmith Court, Station Approach, Broadstone, Poole, Dorset, BH18 8PW. Tel (01202) 659910.
  Monthly periodical with an emphasis on practial construction projects. Also includes regular information on the amateur radio satellite scene.
- *Satellite News*, Edited by Geoffrey Falworth. Details from him at 15 Whitefield Road, Penwortham, Preston, PR1 0XJ.
  A monthly periodical containing detailed information on satellite launches, payloads and operations. A fortnightly bulletin is also available, as is a book list.
- *Short Wave Magazine*, Arrowsmith Court, Station Approach, Broadstone, Poole, Dorset, BH18 8PW. Tel (01202) 659910.
  Monthly periodical which includes sections on both general short wave radio listening and specialist topics. Also includes a regular round-up on the weather satellite scene by the author of this book.
- *Short Wave Magazine* Book Service, Arrowsmith Court, Station Approach, Broadstone, Poole, Dorset, BH18 8PW. Tel (01202) 659930.
  Numerous titles are stocked on a variety of topics associated with short wave radio. Lists are published monthly in *Short Wave Magazine*.
- *Spaceflight* Published by the British Interplanetary Society, 27/29 South Lambeth Road, London, SW8 1SZ. A monthly magazine on general space matters.
- Van Horn, Larry (1987) *Communication Satellites*, Grove Enterprises, PO Box 98, Brasstown, NC 28902, USA.

A book detailing all types of communication satellites, with good coverage of several specialised fields. Comprehensive frequency lists.

Useful addresses

These lists provide useful contact addresses, bulletin boards and computer user groups.

Contact addresses

- American Radio Relay League (ARRL)
  225 Main Street
  Newington
  CT 06111
  USA
- AMSAT-UK
  Ron Broadbent, 94 Herongate Road, Wanstead Park, London NE12 5EQ
  (UK branch of amateur satellite operators)
- The Operations Division
  EUMETSAT
  Am Kavalleriesand 31
  D-64295 Darmstadt
  Germany
- European Space Operations Centre (ESOC)
  5 Robert-Bosch-Strasse
  6100 Darmstadt
  Germany
- National Aeronautics and Space Administration
  Project Operations Branch, Code 513
  Goddard Space Flight Centre
  Greenbelt, MD 20771
- National Space Science Data Center
  Goddard Space Flight Center
  Greenbelt, MD 20771
  Telephone: (301) 286-6695
- RAID BBS access – write to:
  NASA/Goddard Space Flight Center
  Project Operations Branch/513
  Attn: Orbital Information Group
  Greenbelt, MD 20771
- To receive the *NASA Report To Educators* and other NASA

publications, write to the address:
Educational Publications Services
Mail Code XEP
NASA Headquarters
Washington, DC 20546

- Serving inquiries related to space exploration and other activities:
  NASA Jet Propulsion Laboratory
  Teacher Resource Center
  JPL Educational Outreach
  4800 Oak Grove Drive
  Mail Code CS-530
  Pasadena, CA 91109
  Tel (818) 354-6916 Fax: (818) 354-8080
- National Geophysical Data Centre
  E/GC2
  325 Broadway
  Boulder, CO 80303
- National Remote Sensing Centre
  Space Department
  Royal Aerospace Establishment
  Farnborough
  Hampshire
  GU14 6TD
- Remote Imaging Group (RIG)
  Membership secretary:
  Ray Godden
  G4GCE
  Wayfield Cottage
  The Clump
  Chorleywood
  Herts
  WD3 4BG
  Tel (01923) 720714.
  (UK club for weather satellite enthusiasts)
- Space School
  Attn. Rodney Buckland
  Brunel University
  Uxbridge
  Middlesex
  UB8 3PH
- SPACEWARN Bulletin: mailing address
  Chee-ming Wong
  World Data Centre A for Rockets and Satellites

Code 930.2
Goddard Space Flight Centre
Greenbelt
MD 20771
- WEATHERWATCH-UK
RAE Lasham Airfield
ALTON
Hampshire
GU34 5SH
Tel 01256 381 448 – only over weekends or between 1715 and 0845
GMT (weekdays).
(Telephone information about NOAA satellites)

Computer user groups and amateur radio groups

There are numerous user groups set up to cater for almost all interests, and
lists appear regularly in some computer and radio magazines. From time to
time, groups may hold some software of relevance to the satellite community.
The following are a selection:

- Amateur Radio (CBM):
Simon Lewis GM4PLM, Commodore Radio Users Gp, 69 Irvine Drive,
North Clippens, Linwood, Paisley, PA3 3TB.
- Amateur Radio (Comms)
Pat & John Beedie GW6MOJ/MOK, BARTG, Ffynnonias, Salem,
Llandeilo, Wales, SA19 7NP
- Amstrad: several groups, e.g.
1512/1640 PC Independent User Group, The Computer Advice Centre, 87
High Street, Tonbridge, Kent, TN9 1RX. Tel (0732) 771512.
- Atari ST
16/32 PD Library, 35 Northcote Road, Strood, Kent, ME2 2DH.
- BBC Micro
Beebug, 117 Hatfield Road, St Albans, Herts, AL1 4JS
- Commodore – all micros
Jack Cohen, ICPUG, 30 Brancaster Road, Newberry Park, Ilford, IG2
7EP
- Commodore Amiga
UK Amiga Users Group, 144 Charles Street, Leicester, LE2 0QD.
- IBM PC
The IBM PC User Group, PO Box 360, Harrow, HA1 4LQ.
- Scientific PC Users
Tim Bunning, SPCUG, PO Box 17, Retford, Notts, DN22 6BQ

Bulletin boards (BBS)

For those with a computer and modem, there are a number of UK BBS available, where items of interest to the space enthusiast are accessible. I regularly access most of the following BBS. Modem protocol is the standard 8-bit, no parity, 1 stop bit. In alphabetical order:

For those able to dial up American BBS, the number of the *Celestial Bulletin Board*, which I understand carries recent Kepler elements, is [USA] 513-427-0674.

- Dartcom provide a variety of files including Kepler elements, updated once per month. They are on 01822-88249.
- The *Prometheus* BBS includes radio astronomy, rockets and Kepler elements – courtesy AMSAT-UK. To use the system you need a terminal/micro running the Viewdata emulation. The number is 0181-3007177.
- The *Remote Imaging Group* (RIG) BBS carries Kepler elements for the weather satellites, updated weekly, with a good selection of programs and files available for members to download. It is on 01945-440666.
- *Starbase*1 is a BBS devoted to astronomy and space matters. Kepler element files are regularly updated, as are thousands of space-related files, images and NASA publications. Two lines are available: 0171-7033593 and 0171-7016914.
- *Timestep Weather Systems* BBS normally contains the latest Kepler elements for the WXSATs. It is on 01440-820002.

Appendix 2   Glossary of terms

The following list includes many of the terms and abbreviations used in this
book. In several cases a detailed explanation of the terms has already been
given in a particular chapter, so a reference is made to that chapter.

AM	Amplitude modulation; a carrier signal of fixed frequency is modulated (modified) by changing its instantaneous size (amplitude), by the signal to be monitored
AOS	Acquisition of signal; the time that we hear the satel-lite – may be different from the calculated time
Apogee	The part of a satellite's orbit that is its furthest distance from the Earth (see Appendix 3)
a.p.t	Automatic picture transmission (see Chapter 3)
Arg P	Argument of Perigee; the angle (measured at the centre of the Earth) from the ascending node to the perigee (see Appendix 3)
Asc N	Ascending node (see Appendix 3)
AVHRR	Advanced very high resolution radiometer
BBS	Bulletin board service
Checksum	A number, often included in part of a transmission, which provides an indication of the transmission accuracy of the received data. The checksum depends on the actual data itself
CGA	Colour graphics adapter
Clarke Belt	The geostationary orbit, some 35800 km distant, where a satellite has a period of 24 hours
DCP	Data collection platform
Decay	Satellite decay (see Appendix 3)
Descending node:	See Appendix 3
DOS	Disk operating system
Eccentricity	The measure of the circularity (ellipticity) of an orbit (see Appendix 3)

Elevation	Angular height above the local horizon
EMS	Expanded memory specification
Epoch	The precise time to which the parameters of a satellite's orbit refer (see Appendix 3)
ESA	European Space Agency
ESSA	Environmental Science Service Administration
FAX	Facsimile
FM	Frequency Modulation; the process by which a carrier signal of fixed frequency is changed (modulated) by the frequency of another signal
Geostationary satellite:	One (in the Clarke belt) whose orbital period is 24 hours
GHz	Gigahertz (1000 MHz)
GOES	Geostationary Operational Environmental Satellite
GOMS	Geostationary Operational Meteorological Satellite
h.r.p.t.	High-resolution picture transmission
Hz	Hertz; one cycle per second
Inclination	The angle between the plane of a satellite's orbit and the plane of the Earth's equator (see Appendix 3)
i.r.	Infra-red
kHz	Kilohertz (1000 Hz)
LOS	Loss of signal
Mean anomaly	See Appendix 3
Mean motion (MM)	See Appendix 3
MHz	Megahertz (1 000 000 Hz)
NASA	National Aeronautics and Space Administration
NOAA	National Oceanographic and Atmospheric Administration
Node	See Appendix 3
PDUS	Primary data user station (METEOSAT)
Perigee	See Appendix 3
Period	The time for the satellite to make one complete orbit of the earth (see Appendix 3)
Polar orbit	One having an inclination near 90° which therefore takes it near the poles on every orbit.
Polarization	Satellites are often stabilized in orbit by spinning on one axis. This causes the transmission to be polarized, i.e. to have the energy concentrated in a particular direction
SAR	Synthetic aperture radar
Sub-satellite point	The point immediately below a satellite
Sun-synchronous	An orbit (e.g. NOAA) which keeps its orbital plane always facing the Sun
SWL	Short wave listener

TBUS	a.p.t. predictions message
Telemetry	The radio signal transmitted by a satellite
Transponder	The unit which receives an uplink signal from a ground station and re-transmits it to ground, possibly at a different frequency
VGA	Video graphics array
WEFAX	Weather facsimile
XMS	Extended memory specification

Appendix 3   Keplerian elements

Kepler's laws of planetary motion are named after Johannes Kepler, who based the original laws on observations of the planets – the Sun's natural satellites. The laws describe the behaviour of any satellite of any planet, and allow predictions to be made for any future time (epoch), but in the case of artificial Earth satellites, accuracy is limited, depending on various factors.

In order to have a computer program predict when various satellites will be above the horizon at your ground station you have to provide a set of numbers (parameters or elements) for the satellite. Let's look at each in turn.

There are several of these parameters and different computer programs may require slightly different versions! Some programs require the 'equator crossing time'. It is possible to use a set of Kepler elements to calculate other parameters used in different programs.

The first Kepler elements

The first satellite, Sputnik 1, was put into a fairly low orbit above the Earth and had a revolution (orbital) period of little more than an hour. This orbital period is related to its average height above the Earth – the higher the orbit, the longer that period, and therefore the slower its speed. So the Moon, which of course is the Earth's largest natural satellite, takes about one month to orbit the Earth, at its distance of some quarter of a million miles. It is an interesting project to calculate the various orbital speeds for satellite orbits of different heights.

Parameters

Epoch

This element refers to the time at which the Kepler elements were measured. Measurements are usually made by radar but can be derived from optical measurements. They are given in the form: 95 12.2206774 or 01/12/95 05:17:46UTC.

You can see that the first form uses the 'Day of the Year' format in which 1 January is Day 1 and 1 February is Day 32. By multiplying the decimal number by the appropriate factor, you can convert the decimal to the actual time of day – so 0.2206774 (times 24) is really 05 hours and 17 minutes UTC. Remember that sometimes the American date format is used in which the month and date are reversed (as above)!

Orbital inclination

This is the angle between the plane of the satellite's orbit and the plane of the Earth's equator. A satellite in an orbit with an inclination of 0° is travelling in the Earth's equatorial plane – geostationary satellites normally have such inclinations. Orbital inclinations near 90° indicate that the satellite passes over both poles on every pass, while the Earth rotates below. Weather satellites have inclinations near 90° so they pass over every place on Earth. Inclinations can vary between 0° and 180° – those above 90° simply refer to satellites orbiting in the opposite direction. Intermediate inclinations between 10° and 80° are commonly used.

The Earth's diameter is about 12800 km; the satellite's height may be just 800 km above the surface.

Orbital period (mean motion)

The time taken by a satellite to complete one revolution of the Earth is termed its orbital period, commonly about 100 minutes. Geostationary satellites naturally have periods of 24 hours (23 hours 56 minutes to be more accurate). From this period, the number of orbits per day (mean motion – MM) can be calculated. Many of the weather satellites have MM values of about 13 or 14, that being the number of orbits of the Earth that they complete each day. METEOSAT has an MM of 1.00.

The term 'nodal period' is often used in computer programs, and this is defined as the period taken from one perigee passing to the next. In practice it is similar to the orbital period expressed in minutes.

Right ascension of ascending node (RAAN)

With a name like that, one can understand a reluctance to learn about satellites! Those having an interest in astronomy know that astronomers describe the position of objects (stars and planets etc.) in the sky, using just two parameters – declination (which is the angular height of the object above or below the celestial equator) – and right ascension (RA). RA is a straightforward measurement. It is the angle between the object and a place in the sky called 'the first point of Aries'. The stars appear to move once around the sky each day as the Earth rotates on its axis; the Sun appears to slowly move in front of these stars. The position occupied by the Sun at the start of spring (around 21 March) is called the vernal equinox – it was once in the constellation of Aries. Because of the changes (precession) of the direction of the polar axis, this 'first point' has moved, but we still use that position on the celestial sphere to mark the origin of right ascension. So what about nodes?

Try to draw a diagram showing a satellite's orbit at an angle of perhaps 82° to the equator: you realize that the orbit can be drawn in an infinite number of positions – anywhere around the Earth. Join a line between the two places where the orbit crosses (intersects) the equator and we have two locations – called the nodes. We are getting closer!

Think about the satellite's movement around this orbit; at one node it crosses the equator going north (ascending!) and the other while going south (descending). We always use the ascending node for Kepler elements.

Eccentricity

Orbits of satellites are never simple circles; they are some variation of an ellipse. Technically a satellite can have any of several differently shaped orbits, but the common ones vary between an elongated ellipse and a near circle. Eccentricity is the measure of this difference. A circle has an eccentricity (e) of 0; an elongated ellipse has an e approaching 1. Eccentricities therefore lie between 0 and 1. One 'diameter' of the orbit is normally longer than the other, so at one part of the orbit the satellite comes nearest to the Earth – this point is called the perigee. The point furthest from the Earth is called the apogee.

Argument of perigee

Previous parameters now allow us to draw the orbit fairly accurately, but we don't yet know how to position the elliptical orbit around the Earth. Where does the long axis of the ellipse point?

Imagine a line drawn between the apogee and the perigee – this is the long diameter (or major-axis) of the ellipse. The argument of perigee is the angle (measured from the centre of the Earth) from the ascending node to the perigee, previously described. It varies between 0° and 360°. The word argument is a mathematical one meaning a parameter – in this case, an angle.

Mean anomaly

We have almost finished defining the parameters that position our satellite in its orbit. All the previous measurements enable us to draw the actual orbit around the Earth. Where is the satellite currently in that orbit? Our reference point is the perigee and our reference time is the epoch. Starting at 0° at perigee we measure along the orbital ellipse until we reach the satellite, and this angle is called the anomaly.

Drag

This parameter (sometimes called acceleration) is not always given with Kepler elements, and its use is somewhat arbitrary. The Earth's atmosphere affects the orbits of satellites passing through – usually satellites orbiting below several hundred kilometres. Geostationary and other distant orbiting satellites are affected differently! The drag parameter refers to the change (increase or decrease) in mean motion of the satellite.

The orbits of near-Earth orbiting satellites tend to decrease in size, that is, tend to decay. Orbits approaching decay have an MM of something over 16 revolutions per day. This means that the atmosphere is actually causing the satellite to speed up as its height decreases. Hence the drag has a positive value.

Consequently, low orbiting satellites, such as the Russian MIR complex and the Shuttle, have relatively high drag parameters, whereas the Russian METEOR class 3 satellites have a relatively low drag.

Technically, drag is the rate of change of the number of revolutions per day. When it is missing from an element set, you can improve longer-term prediction accuracy by using a figure of about 0.000001.

Predictions

Having input your Kepler elements into a suitable computer program, assuming that they are recent elements, you can expect the resulting predictions to be reasonably accurate. They should remain so for a few weeks. Unfortunately several things conspire to cause the predictions to become less and less accurate with time!

Firstly, the computer program usually makes certain assumptions about the satellite's orbit, using methods of approximation which themselves vary in accuracy.

Secondly, the drag parameter, assuming one was given, actually varies with time. As explained previously, drag is primarily caused by the upper atmosphere, which has a density which varies continuously. Drag is therefore, at best, an approximation. Solar activity affects it, and has a larger effect on lower orbiting satellites than those higher up.

If you monitor Kepler elements for a time, you will see how wildly the drag can change. During sunspot minimum it may remain constant for a given satellite, assuming that it is in a circular orbit.

Consequently, the higher the orbit, the longer, in general, will the predictions remain accurate. Lower orbiting satellites, such as OKEAN-3, suffer a larger drag than the NOAAs, which in turn have a larger drag than METEOR class 2 satellites, which in turn have a larger drag than METEOR class 3 satellites. Sometimes the drag may be negative! This is because on some occasions, the gravitational attraction of the more distant Moon may act to increase the satellite's orbital period instead of decreasing it.

Similarly, geostationary satellites have exceedingly low drags which may occasionally become negative.

Deriving Kepler elements

The first satellite that I identified as being 'new' was METEOR 2-16. This was followed by logging METEORS 2-17, 2-18, 3-2, 3-3, 3-4, 3-5 and more recently, 3-6. On each occasion I carefully logged AOS and LOS times, making estimates where necessary. With experience, I think one can judge the maximum elevation reached by the satellite to within about 10°, and, of course, the actual direction of travel – whether north- or southbound.

Ideally after that 'first pass' we would ring up NORAD and request Kepler elements, and if they were monitoring the satellites as well as we are, they could let us have a new set! Of course life isn't like that; when we hear a new 'bird' – be it a METEOR or COSMOS, we are on our own. So I sat down with my notes and thought about the problem.

From your first observations of the picture from any new satellite, you need to identify the class to which it belongs – NOAA, METEOR 2 or 3, or COSMOS satellites. Launches of NOAA satellites are scheduled in advance, and the frequencies are known. COSMOS satellites are much more difficult to analyse because their transmissions are not normally continuous, so even with an official element set you may not hear the satellite after two orbits have passed! It is therefore the METEORS that are worth pursuing.

Any new METEOR should be class 3 category; this immediately fixes the eccentricity and orbital period (height). Let's assume that we have noted AOS, LOS, maximum elevation and direction times for the new satellite – believed to be a class 3 METEOR. Next we look at some elements from a comparable satellite – say METEOR 3-6. If these predictions are run for the day in question then a set of passes is obtained which are now compared with the times of the new satellite.

The next step involves putting 'dummy' elements for your new satellite, in your predictions program. You can use a comparable satellite and modify its parameters, and then run the software to compare the new predictions with your actual observations.

The first item to compare is the direction of the satellite's travel, compared with the satellite's known movements at that time. It may be completely opposite for the specified time period. The first parameter to change is the RAAN value. By adjusting this, you will come across a figure that produces AOS and LOS times that are close to your measured ones, and in the correct direction, e.g. southbound. This process may take 5 or 50 minutes! During it, you will see the times change in sequences until they approach quite closely. Then change the MA. By this re-iterative process you should approach a workable 'dummy' set of parameters.

Within a few days you will probably be able to get a measured set of Kepler elements, but meanwhile you should find your derived set help in predicting when to listen.

Appendix 4   Product reviews

JVFAX7.0

Eberhard Backeshoff is the author of a comprehensive decoding program called JVFAX, which, at the end of 1994, was in its seventh version. He has made the program available to the satellite community without charge. The software requires an interface unit, and can then decode a variety of utility FAX, as well as weather satellite telemetry.

For easy use of software and interface, it is essential to configure the system correctly; time must be spent adjusting signal levels and software parameters. JVFAX is written for the PC, and I tested it on a 486SX running at 25 MHz. It will run on slower machines, particularly if you use an 'active' interface – one doing some signal processing before passing data to the computer.

Settings

I connected the unit to my mouse's port (COM1), which has an address (location in the computer's RAM) of 03f8h; this figure should be checked during configuration of JVFAX. Instead of unplugging your mouse, you may prefer to use COM2 (the second serial port – if fitted), in which case the address needs changing to 02f8h.

Using live telemetry from a NOAA WXSAT, the program first displays the incoming signal. During reception, JVFAX shows the black-to-white spectrum content in a miniature screen display. When a.p.t. is detected, the program screen display acknowledges presence of the format. The signal from my receiver initially swamped the interface, so to improve this balance between receiver and interface, I removed the cover from my receiver, and, after locating my notes (which date back to the mid-1980s), I reduced the output signal level from the receiver – obtaining a sensible spread of black and white in the signal.

The on-screen picture should show a complete grey scale being received.

Check grey scale included in METEOR transmissions, and NOAA calibrations, each of which includes several levels. To confirm all is well, 'Quit' the program, then re-start it. An automatic triggering into 'Squelch on' shows that adjustment is nearly complete. Using real-time signals from NOAA and METEOR WXSATs, I checked that triggering was accurate in each case, the result was a clear, well-defined picture from each WXSAT. Signal strengths differ between METEORs and NOAAs but some compensation can be made within the software.

Picture alignment and synchronization

A poorly aligned picture has a non-vertical edge – tilting in one direction. Doppler shift on the satellite signal may produce a tilted image, but there is provision for compensation within JVFAX. Initial adjustment is made via the '/' keyboard character; a line appears on the screen, and slope adjustment allows correction of the new image. A second adjustment should complete this correction, producing a straight edge. The use of the 'roll' option can reposition the picture edge in the correct place. These adjustments are software related, rather than being a function of the hardware (interface) unit, so full setting up of the combined system is necessary to obtain the highest quality images.

The program can give a virtually perfect image. JVFAX permits the attainment of 256 grey levels, if you have sufficient memory available. Using the 64 grey level setting, I monitored several METEOR 3-5 passes, and looked carefully at the picture quality. The software operates slightly differently from some other systems – it uses picture scroll – so you can see most of the detail as the pass proceeds.

With hardware adjustments made (and fixed), a typical METEOR pass will show land as dark areas, so you may wish to enhance images post-pass. This is done without difficulty, using the enhance feature in edit mode – but again, it is software related, rather than a function of the interface unit.

PCGOES/WEFAX (PC WeatherSAT)

PCGOES is PC software designed to decode both satellite a.p.t. telemetry and FAX signals, when connected to a suitable receiver. The interface is a 25-way connector for plugging into the serial port at the back of your PC, and includes input cables for the audio signals from your weather satellite receiver and from your short wave receiver.

The computer needs a minimum of 640 kb RAM to provide 640 × 800 pixels with 16 grey levels on a VGA monitor (or better). It can use extra memory to provide higher-quality pictures.

An excellent book, written by John E. Hoot of Software Systems Consulting

(USA), describes the software and installation, and a cassette tape of sample FAX data and 3.5 in or 5.25 in disk containing the software is also included. The book gives considerable information about weather satellites and lists FAX transmission frequencies in a comprehensive series of lists.

The program is menu driven, and after setting up, there are sample pictures for trying the image enhancement facilities. A simulated oscilloscope screen lets you optimize decoding of the input signal, and has an automatic option.

The program includes printing hard copy, basic animation, and programmed keys for specific jobs, such as image reversal, are useful facilities. The major missing option is provision of north to south picture display. After a pass, the image can be instantly reversed.

When an image first appears it may not be correctly synchronized; a key press causes synchronization. Saved images can be adjusted after capture. When the screen has filled with the image, it scrolls, but when the RAM buffer holding this image has filled, the screen freezes.

METEOSAT images can be programmed, collected and displayed sequentially, and although presented upside-down, they are reversed automatically before saving. Software includes an option to produce animated sequences from METEOSAT; I found that the best results were from whole images.

Picture synchronization is provided by the computer's own clock, and software allows the scanning rate to be set for any live satellite signal, so you are guaranteed a synchronized picture. Doppler changes, caused by the satellite's motion, may cause the picture to tilt during the pass, but I did not find this to be a problem, and the pictures stayed locked even during temporary signal loss, which occurs near AOS and LOS.

There is a comprehensive section to predict satellite tracks and to update the orbital elements – an extremely important facility. This section can help you identify which satellites are in use, and there is a map display.

FAX

FAX is covered, in addition to satellite pictures, and automatic capture works well. Image enhancement options include contrast adjustment and false colour, for which several sample palettes are included. You can also produce 'true' colour images using the 'press images' option, in which FAX pictures in cyan, magenta and yellow are received, then processed.

The printing facilities included with this program are good, and can be done from zoom mode. You can also export files to desk-top publishing programs.

PROSAT 2

This program suite comes with hardware to fit an expansion slot inside your PC. A '286' or better computer, with VGA/SVGA monitor, is required. The software can provide high-quality images from a.p.t. transmissions. It is installed on the hard disk; some 5 Mb should be available because the software can store the whole of the longest passes – files up to 2.4 Mb.

The complete suite comprises four parts; VGASAT4 for processing METEOSAT data, MEGANOAA for processing all polar orbiting satellites, ANIMATE for animating METEOSAT frames, and Track II for satellite predictions.

Each section has options catering for most conceivable requirements. MEGANOAA displays live imagery, storing it on hard disk if your computer has insufficient RAM. Options allow satellite selection and synchronization as appropriate. METEOR satellites include a choice between *unsynchronous* for class 2 satellites and *synchronous* for class 3 METEORS. For NOAA you can have *line-by-line* or *start-only* synchronization. For OKEAN there is a larger choice, with the promise of updates should any new format be required.

Image processing options are good, with preset or user-designed palettes for colour. Images can be examined closely and a portion transferred to the METEOSAT section for further processing. NOAA passes utilize thermal calibrations so you can measure the temperature of any selected point. There is a temperature slice option showing areas of equal temperature with a selected colour. This is superb and can be used for detailed seasonal temperature studies.

METEOSAT frame collection is programmable and can store as many as your disk can accommodate – at 512 kb each frame! Colour facilities are good. Numerous processing techniques are available – country outline removal and noise reduction – if your pictures contain excessive noise.

ANIMATE allows selection of a portion of a frame for producing a series of images. RAM can be used for image storage – 4Mb can store 29 images, or the hard disk can be used instead and stored to its limit. Animation speed can be adjusted; image collection set, with individual names for each series; some image enhancement can be programmed before collection, and the number of images per set is adjustable.

Available from Timestep Weather Systems Ltd

INSTANT-TRACK

This satellite tracking program has numerous facilities. Written by Franklin Antonio, it runs on a PC which, if fitted with a VGA monitor, provides good graphics. The PC needs a minimum of 512 kb of memory.

The menu gives a choice of options including real-time text display of the satellite's path, a map display, an ephemeris, visibility schedule, and provision for updating the orbital elements. Some options have other choices within. Satellite selection is from the database holding up to 200 satellites, and there is a choice of map projection. The Mercator projection shows the world with the satellite's path shown as a circle representing the footprint.

In tracking mode, several keys provide further facilities:

'f'   goes to fast forward mode (and returns to normal)
'p'   gives a 'bird's eye view' as seen from the satellite and repeating 'p'
          shows the orbital position around the Earth; the starry background is
          the final 'p' option.
'w'   shows when the next rise or set time will be
't'   allows selection of any date/time to be made
's'   toggles to a map scroll
'o'   adds a selected observer

The update option allows the use of either AMSAT-type Kepler elements, or
NASA two-line elements, to be input automatically. If you have access to a
bulletin board to download these NASA elements it is a doddle!

Having selected Mercator display for any satellite, you can switch to the
next satellite simply by pressing the right cursor key. Grouping the various
satellites together in the database is a good idea. Putting the weather satellites
together, then amateur radio satellites, navigation ones etc., it is very easy to
switch between them.

Checking orbital elements can be done in tracking mode – pressing 'e' shows
the element list; 'd' shows derived values, such as perigee height etc., and the
cursor keys let you change satellites.

Your station parameters can be entered, as can other stations, and there are
over 1700 cities in the database. The software includes a comprehensive 'help'
facility and comes with a good manual.
Available from Timestep Weather Systems Ltd.

PCTrack version 3.1

A copy of PCTrack, written by Thomas C. Johnson of Johnson Scientific
International, was downloaded from a Bulletin Board. A PC-compatible
computer using an 80286, 80386(SX or DX) or 80486(SX or DX) CPU is
needed, running DOS version 5.0 or later, and a minimum of 2 Mb RAM. A
maths co-processor is preferred but is not essential, and about 5 Mb of hard
disk space are required. A mouse is convenient during initial setting up. The
program runs in DOS protected mode, so can access all system memory, and
uses standard VGA (640 by 480 pixels).

General description

PCTrack is shareware; the author asks users to register – cost $45 (about
£30) – for which they receive an enhanced version. The software comes as a

suite of files and programs, easily installed on a hard drive. It includes two large databases holding up to 300 satellites and 300 sites. Calculations are done to double precision accuracy, so for the best results, recent satellite Kepler elements should be used. Several satellites can be shown simultaneously on the graphical display, including footprints (circles of visibility on the earth below), and groups of satellites (weather, amateur radio satellites, Glonass) can be set up.

Many parameters – colours associated with certain items, sites, time formats and views – are adjustable, giving considerable control over the display. Future passes can be identified, and even sections of the pass which are illuminated by the Sun can be seen – ideal for monitoring 'visible' satellites. The updating of Kepler elements is almost automatic.

Getting started

Installation is straightforward, using the program provided. It came with a full database of satellites and had several sites (major cities around the world) pre-programmed – I edited in Plymouth. The program is started by entering PCT.

The starting screen contains a Menu – choices include *File, Edit, View* and *Options*, and are selected either by clicking with a mouse or using F10 to activate, then pressing the first letter – *F, E, V* or *O*. Mouse operation is considerably more convenient. Some sections require entries in various parts of the screen – the mouse was speedier than the Tab key for these entries.

The sensible choice for the first use of the program is to configure the software using *Options, System configure*. This allows setting of the time/date formats – I set time to UTC, with no offset – then set the date format to the English standard – date:month:year.

Several function keys are programmed, and act differently according to the screen display; for example, when used from the Main Menu, F3 starts text display tracking of the primary satellite, but selects global viewpoint when used from the tracking display. With the built-in help display there were no problems. From the Main Menu, the program can be started in graphical tracking mode by pressing F2, or selecting *View, Track Graphics*.

Parameter files

From the Main Menu, function key F2 loads the default parameter file – 'default.prm', or whichever file was last used. This file, and others already present (called *.prm), has been pre-set with satellites and American sites. The screen shows either a three-dimensional globe with orbiting satellites, or a Mercator projection, depending on which was previously used. Press F4 to swap between 3D-global view and Mercator.

PCTrack allows the user to generate different satellite groups and to associate different sites with each. Each grouping is called a parameter file and the satellites selected must already exist in the Master satellite database. From this, individual files having any combination of satellites can be saved as parameter files. To access the file-editing facility, we return to the initial Main Menu screen by pressing *Escape*.

Using *File, Open* you are shown the pre-set parameter files, e.g. weather.prm, any of which can be edited. I prepared a METEORS.prm file, to include all current group three METEORs together with later group two METEORs – totalling eight satellites. I also modified the weather.prm file; nearly all the CIS WXSATs were already present in the database.

Editing files

To customize your satellite parameter file, you *Open* it from the Main Menu, then select *Edit, Satellites* (or press F5). The display allows editing of the Master Satellite database; you can *Edit, Add, Delete* or *Activate* a selected satellite. This is also where you select your Primary satellite – the one which runs in text tracking mode.

This facility is intuitive; when *Active* is selected, the Master database is on the left of the screen, the right side shows currently active satellites selected for the parameter file. With a click of the mouse you select a satellite for transfer to or from the active set, or delete it from the Master database. You can create a new parameter file by saving the present one with a new name.

Sites which appear on the display are called the primary and secondary sites, though only one need be selected. They are edited by selecting F6 or *Edit, Sites* from the Main Menu. Clicking the mouse on the site, followed by transferring it to the Primary site box, is all that is required. You can edit a site by entering your latitude and longitude, as accurately as you can obtain. Atlases or a local Ordnance Survey map can provide this information, and help you estimate your height above sea level.

Importing satellites

Updating the database with new Kepler elements is straightforward. Having new NASA two-line elements called 'apt.all', in a directory called Kepler, they are imported using *File, Import*, and can read either 2-line or AMSAT format. The full filename and path is entered and the file read. New satellites can be entered in the same way – assuming the database is not already full.

Changing the display

Within the graphical display screen, a new Menu of hidden keys is available: to see the list, press F1 (the Help key). Within tracking mode, using F2 you can activate a fast-forwards display; this mode calculates where the satellites are going. The amount by which time is incremented in fast-forwards mode is adjustable using the left/right cursor keys to select the second, minute or hour parameter; the up/down keys then increment and decrement respectively.

Both displays – Mercator and 3D – are impressive, and, as becomes increasingly clear with this program, both are adjustable. While in global view (F4 switches to global from Mercator), press F3 to change the position of your viewpoint. There are eight choices of perspective, four fixed longitudes seen from near the globe, and four from further away. Each option also illustrates the satellite's height above the globe. I wanted to change the viewpoint to zero longitude, rather than remain over the USA; documentation explains that the registered version allows this flexibility.

On first running this program, the tracking display included two other sections, one showing the primary ground site selected for this file. Below this was an AER chart – azimuth, elevation and rising/setting of each satellite when above the horizon of the selected site. You can remove or display the ground site graphics using F6. Without the sites, the Mercator map or globe expands to fill the screen. To see the AER display in action (assuming it was empty while you were running the program), press F2 to activate fast-forwards. As each satellite comes over the horizon, its entry appears in the lower section, showing its calculated position. Tracking can be paused using F7. After changing certain options you may want to refresh the screen – press F5.

Attributes

The versatility of this software becomes apparent as each new option is tried. Parameter files have attributes which can be changed. To select any attribute, you must use the tab key, then change it using the up/down cursor keys. The mouse is not used in this pull-down window. Attributes include footprints, satellite name display, sun position, site names and satellite lines of sight (lines connecting the ground station to the satellite, when above the horizon – abbreviated to LOS). Several attributes have multiple choices, e.g. footprints; this attribute can have any of five options – single (no trail is left), cont (continuous, in which the satellite leaves a trail), svis (the attribute is only shown when the satellite is visible from the site, and it leaves no trail), cvis (as for svis, but a trail is left during the period of site visibility), and off (no footprint). Setting either cvis or svis and enabling fast-forwards, quickly clutters up the screen with satellite trails – hence the use of F5 to clear the display.

Certain display colours can be changed using F9. Care has to be taken while doing this.

Printouts

The software includes a facility to print pass details for future reference. A range of printers can be configured. Using the *Scan* option, a time period can be entered and a list of mutually visible passes for the ground stations will be produced. As usual there is a choice! You can produce a disk file (for incorporation into a word processor) or simply print it out directly.

Documentation

Several files are included with the program suite; a quick-look file called QUICKSTRT.DOC (16 kb) and a larger tome called USRGUIDE.DOC, shows just how comprehensive this software is. The index alone is longer than documentation supplied with some commercial software!

Problems and bugs

If you do not have a co-pro, a 486DX or a reasonably fast machine, you may find the program updates slowly, possibly taking several seconds between updates. The software calculates the positions of all active satellites in the current parameter file – the more satellites on display, the longer it takes. This didn't bother me, except when making changes – the software only responds after completing its calculation sweep – so one can feel impatient when there are a large number of active satellites. In normal operation, with eight satellites displayed, on a slow machine it may momentarily appear to have stopped; on my 33 MHz 386DX (with co-pro) screen updates occurred every 1.5 seconds. My 25 MHz 486SX took 12 seconds between screen updates. With two satellites displayed, updates then took about 0.5 second. This is not a meaningful problem when you know what is happening. Running on a 486DX2-66 MHz machine the update took less than 1 second.

The screen display is VGA, using 640 by 480 pixels, a little low by today's standards, though generally acceptable.

Bugs? I have not found any so far!

Availability

If you are a member of the Remote Imaging Group, PC-Track can be downloaded from their BBS on (01945) 440666. You may also find it on other specialized BBS. Alternatively, send me a (PC-compatible) 3.5 in disk, together with pre-paid return package and 50p towards collection costs and I will provide a copy by return, including the latest Kepler elements. These are also available routinely from me – see the 'Info' column in *Short Wave Magazine*.

Note

These reviews are for products that suppliers have kindly allowed me to examine. Further reviews appear from time to time in my column in *Short Wave Magazine*, from which sections of the above have been used.

Appendix 5 Dissemination schedule METEOSAT 5

Dissemination Schedule S9410M02 - METEOSAT 5 (0 degrees W)   FINAL

(Channel A1 = 1691 MHz - Channel A2 = 1694.5 MHz) Valid from 18 October 1994 (provisional)

HH	00		03		06		09		12		15		18		21		HH
MM	CH A1	CH A2	CH A1	CH A2	CH A1	CH A2	CH A1	CH A2	CH A1	CH A2	CH A1	CH A2	CH A1	CH A2	CH A1	CH A2	MM
2	D1 48	AIW 48	D1 06	AIW 06	C02 12	AIVH 12	C02 18	AIVH 18	C02 24	AIVH 24	C02 30	AIVH 30	D1 36	AIVH 36	D1 42	AIVH 42	2
6	D3 48	AIW 48	D3 06	AIW 06	C03 12	AIVH 12	C03 18	AIVH 18	C03 24	AIVH 24	C03 30	AIVH 30	D3 36	AIVH 36	D3 42	AIVH 42	6
10	D4 48	AIW 48	D4 06	AIW 06	D1 12	AIVH 12	D1 18	AIVH 18	D1 24	AIVH 24	D1 30	AIVH 30	D4 36	AIVH 36	D4 42	AIVH 42	10
14	D5 48	ETOT 06	D5 06	DTOT 06	D3 12	BW 12	D3 18	BW 18	D3 24	BW 24	D3 30	BW 30	D5 36	BW 36	D5 42	DTOT 42	14
18	D6 48	ETOT 48	D6 06	ETOT 06	D4 12	DTOT 12	D4 18	DTOT 18	D4 24	DTOT 24	D4 30	DTOT 30	D6 36	DTOT 36	D6 42	ETOT 42	18
22	D7 48		D7 06		D5 12	ETOT 12	D5 18	ATEST1	D5 24	CTOT 24	D5 30	CTOT 30	D7 36	ETOT 36	D7 42	ATEST1	22
26	D8 48	GMSA 48	D8 06	GMSA 06	D6 12	GMSA 12	D6 18	ATEST1	D6 24	LXI 23	D6 30	GMSA 30	D8 36		D8 42	ATEST1	26
30	D2 01	BIW 1	D2 07	BIW 07	D2 13	BIV 19	D2 19	BIV 19	D2 25	BIV 25	D2 31	BIV 31	D2 37	BIV 37	D2 43	BIW 43	30
34	D9 01	AIW 01	D9 07	AIW 07	C02 13	AIVH 13	C02 19	AIVH 19	C02 25	AIVH 25	C02 31	AIVH 31	D9 36	AIW 37	D9 43	AIW 43	34
38	D1 01	AIW 01	D1 07	AIW 07	C03 13	AIVH 13	C03 19	AIVH 19	C03 25	AIVH 25	C03 31	AIVH 31	D1 37	AIW 37	D1 43	AIW 43	38
42	D3 01	AIW 01	D3 07	AIW 07	C3D 13	AIVH 13	C8D 19	AIVH 19	C3D 25	AIVH 25	C8D 31	AIVH 31	D3 37	AIW 37	D3 43	AIW 43	42
46					C2D 13	AW 13	C2D 19	AW 19	C2D 25	AW 25	C9D 31	AW 31					46
50		GMSB 48		GMSB 06	D3 13	AW 13	C2D 19	AW 19	C1D 25	AW 25	D3 31	AW 31		GMSA 36		GMSA 42	50
54		LXI 01		LXI 07	D1 13	LXI 13	D1 19	LXI 19	D1 25	LXI 26/26	D1 31	LXI 31/32		LXI 37/38		LXI 43/44	54
58	D2 02	BIW 02	D2 08	BIW 08	D2 14	BIV 14	D2 20	BIV 20	D2 26	BIV 26	D2 32	BIV 32	D2 39	BIV 38	D2 44	BIW 44	58

HH	01		04		07		10		13		16		19		22		HH
MM	CH A1	CH A2	CH A1	CH A2	CH A1	CH A2	CH A1	CH A2	CH A1	CH A2	CH A1	CH A2	CH A1	CH A2	CH A1	CH A2	MM
2	D1 02	AIW 02	D1 08	AIW 08	C02 14	AIVH 14	C02 20	AIVH 20	C02 26	AIVH 26	C02 32	AIVH 32	D1 38	AIVH 38	D1 44	AIW 44	2
6	D3 02	AIW 02	D3 08	AIW 08	C03 14	AIVH 14	C03 20	AIVH 20	C03 26	AIVH 26	C03 32	AIVH 32	D3 38	AIVH 38	D3 44	AIW 44	6
10		AIW 02	E1 08	AIW 08	D7 14	AIVH 14	D7 20	AIVH 20	D7 26	AIVH 26	D7 32	AIVH 32	E1 38	AIVH 38		AIW 44	10
14		LY 01	E2 08	LY 07	D8 14	BW 14	D8 20	BW 20	D8 26	BW 26	D8 32	BW 32	E2 38	BW 37		LY 43	14
18		LR 01	E3 08	LR 07	D1 14	LY 13	D9 20	LY 19	D9 26	LY 25	D9 32	LY 31	E3 38	LY 37		LR 43	18
22	GMSC 48		E4 08	GMSC 06	D3 14	LR 13	D3 20	LR 19	D3 26	LR 26	D3 32	LR 31	E4 38	LR 37		GMSC 42	22
26	GMSD 48		E5 08	GMSD 06		GMSB 12		GMSA 18		LZ 25		LZ 31	E5 38	LZ		GMSC 42	26
30	D2 03	BIW 03	D2 09	BIW 09	D2 15	BIV 15	D2 21	BIV 21	D2 27	BIV 27	D2 33	BIV 33	D2 39	BIV 39	D2 45	BIW 45	30
34	D1 03	AIW 03	D1 09	AIW 09	C02 15	AIVH 15	C02 21	AIVH 21	C02 27	AIVH 27	C02 33	AIVH 33	D1 39	AIVH 39	D1 45	AIW 45	34
38	D3 03	AIW 03	D3 09	AIW 09	C03 15	AIVH 15	C03 21	AIVH 21	C03 27	AIVH 27	C03 33	AIVH 33	D3 39	AIVH 39	D3 45	AIW 45	38
42		AIW 03	E6 09	AIW 09	D1 15	AIVH 15	D1 21	AIVH 21	D1 27	AIVH 27	D1 33	AIVH 33	E6 39	AIVH 39		AIW 45	42
46			E7 09	AIW 09	D3 15	GMSC 15	C1D 21	GMSB 21	D3 27	GMSA 24	D3 33	GMSB 30	E7 39	GMSB 36		GMSD 42	46
50			E8 09			GMSD 12	C1D 21	GMSC 18	C1D 27	GMSB 24	C1D 33	GMSC 30	E8 39	GMSC 36			50
54		LXI 03	E9 09	LXI 09	C2D 15	LXI 15	C2D 21	LXI 21	C2D 27	LXI 27/28	C2D 33	LXI 33/34	E9 39	LXI 39/40		LXI 45/46	54
58	D2 04	BIW 04	D2 10	BIW 10	D2 16	BIV 16	D2 22	BIV 22	D2 28	BIV 28	D2 34	BIV 34	D2 40	BIW 40	D2 46	BIW 46	58

HH	02		05		08		11		14		17		20		23		HH
MM	CH A1	CH A2	CH A1	CH A2	CH A1	CH A2	CH A1	CH A2	CH A1	CH A2	CH A1	CH A2	CH A1	CH A2	CH A1	CH A2	MM
2	D1 04	AIW 04	D1 10	AIW 10	C02 16	AIVH 16	C02 22	AIVH 22	C02 28	AIVH 28	C02 34	AIVH 34	D1 40	AIW 40	D1 46	AIW 46	2
6	D3 04	AIW 04	D3 10	AIW 10	C03 16	AIVH 16	C03 22	AIVH 22	C03 28	AIVH 28	C03 34	AIVH 34	D3 40	AIW 40	D3 46	AIW 46	6
10	TEST	AIW 04	ADMIN	AIW 10	C3D 16	AIVH 16	C3D 22	AIVH 22	C3D 28	AIVH 28	C1D 34	AIVH 34	TEST	AIW 40	ADMIN	AIW 46	10
14		ADMIN		TEST	TEST	BW 16	C4D 22	BW 22	C4D 28	BW 28	C4D 34	BW 34		GMSD 36		TEST	14
18							ADMIN	GMSD 18	TEST	GMSC 24	ADMIN	GMSD 30					18
22																	22
26																	26
30	D2 06	BIW 06	D2 11	BIW 11	D2 17	BIV 17	D2 23	BIV 23	D2 29	BIV 29	D2 36	BIV 36	D2 41	BIW 41	D2 47	BIW 47	30
34	D1 06	AIW 06	D1 11	AIW 11	C02 17	AIVH 17	C02 23	AV 23	C02 29	AIVH 29	D1 36	AIVH 36	D1 41	AIVH 41	D1 47	AIW 47	34
38	D3 06	AIW 06	D3 11	AIW 11	C03 17	AIVH 17	C03 23	AV 23	C03 29	AIVH 29	D3 36	AIVH 36	D3 41	AIW 41	D3 47	AIW 47	38
42		AIW 06	E1 11	AIW 11	C5D 17	AIVH 17	C5D 23	AV 23	C5D 29	AIVH 29		AIVH 36		AIW 41	E1 47	AIW 47	42
46		ATEST2	E2 11		C6D 17	ATEST2	E2 23	AV 23	C6D 29	ATEST2	E2 36	TEST		ATEST2	E2 47		46
50		ATEST2	E3 11		C7D 17	ATEST2	E3 23	AV 23	C7D 29	ATEST2	E3 36	ADMIN		ATEST2	E3 47		50
54	CTH 04	LXI 06		LXI 11	CTH 16	LXI 17		AV 23	CTH 29	LXI 29/30		LXI 35/36	CTH		LXI 41/42		LXI 47/48	54
58	D2 08	BIW 08	D2 12	BIW 12	D2 18	BIV 18	D2 24	BIV 24	D2 30	BIV 30	D2 36	BIV 36	D2 42	BIW 42	D2 48	BIW 48	58

AIVH	HRI Full Disc IR & Half Res VIS	BIW	HRI European sector IR & WV
AW	HRI Full Disc WV	BIV	HRI European sector IR & Full Res VIS
AV	HRI Full Disc Full Res VIS	BIVW	HRI European sector IR, WV & Half Res VIS
AIW	HRI Full Disc IR & WV	BW	HRI European sector WV
Lxx	Meteosat X-ADC relay transmission	ATEST1	HRI Test Pattern
	(uplinked by CMS Lannion)	ATEST2	HRI Encrypted Test Pattern
LXI	HRI format (odd slots contain IR data, even slots VIS, when available)		
LZ,LR,LY	WEFAX formats		
GMSx	GMS WEFAX relay transmission (uplinked by CMS Lannion, 5 bit resolution)		

Cnn	WEFAX VIS Full Res
CnD	WEFAX VIS Half Res
Dn	WEFAX IR
En	WEFAX WV
CTH	WEFAX processed Cloud Top Height
CTOT	WEFAX Full Disc VIS
DTOT	WEFAX Full Disc IR
ETOT	WEFAX Full Disc WV
ADMIN	WEFAX Administration Message
TEST	WEFAX Test Pattern

Index